Crop Production Scienc

Series Editor: Jeff Atherton, Professor of Tropical Horticulture, University of the West Indies, Barbados

This series examines economically important horticultural crops selected from the major production systems in temperate, subtropical and tropical climatic areas. Systems represented range from open field and plantation sites to protected plastic and glass houses, growing rooms and laboratories. Emphasis is placed on the scientific principles underlying crop production practices rather than on providing empirical recipes for uncritical acceptance. Scientific understanding provides the key to both reasoned choice of practice and the solution of future problems.

Students and staff at universities and colleges throughout the world involved in courses in horticulture, as well as in agriculture, plant science, food science and applied biology at degree, diploma or certificate level will welcome this series as a succinct and readable source of information. The books will also be invaluable to progressive growers, advisers and end-product users requiring an authoritative, but brief, scientific introduction to particular crops or systems. Keen gardeners wishing to understand the scientific basis of recommended practices will also find the series very useful.

The authors are all internationally renowned experts with extensive experience of their subjects. Each volume follows a common format, covering all aspects of production, from background physiology and breeding to propagation and planting, through husbandry and crop protection to harvesting, handling and storage. Selective references are included to direct the reader to further information on specific topics.

Titles Available:
1. **Ornamental Bulbs, Corms and Tubers** A.R. Rees
2. **Citrus** F.S. Davies and L.G. Albrigo
3. **Onions and Other Vegetable Alliums** J.L. Brewster
4. **Ornamental Bedding Plants** A.M. Armitage
5. **Bananas and Plantains** J.C. Robinson
6. **Cucurbits** R.W. Robinson and D.S. Decker-Walters
7. **Tropical Fruits** H.Y. Nakasone and R.E. Paull
8. **Coffee, Cocoa and Tea** K.C. Willson
9. **Lettuce, Endive and Chicory** E.J. Ryder
10. **Carrots and Related Vegetable** *Umbelliferae* V.E. Rubatzky, C.F. Quiros and P.W. Simon
11. **Strawberries** J.F. Hancock
12. **Peppers: Vegetable and Spice Capsicums** P.W. Bosland and E.J. Votava
13. **Tomatoes** E. Heuvelink
14. **Vegetable Brassicas and Related Crucifers** G. Dixon
15. **Onions and Other Vegetable Alliums, 2nd Edition** J.L. Brewster
16. **Grapes** G.L. Creasy and L.L. Creasy
17. **Tropical Root and Tuber Crops: Cassava, Sweet Potato, Yams and Aroids** V. Lebot
18. **Olives** I. Therios
19. **Bananas and Plantains, 2nd Edition** J.C. Robinson and V. Galán Saúco
20. **Tropical Fruits, 2nd Edition Volume 1** R.E. Paull and O. Duarte
21. **Blueberries** J. Retamales and J.F. Hancock
22. **Peppers: Vegetable and Spice Capsicums, 2nd Edition** P.W. Bosland and E.J. Votava
23. **Raspberries** R.C. Funt and H.K. Hall

Raspberries

Edited by

Richard C. Funt, PhD

The Ohio State University (emeritus), USA

and

Harvey K. Hall, MS

Shekinah Berries Ltd, New Zealand

CABI is a trading name of CAB International

CABI
Nosworthy Way
Wallingford
Oxfordshire, OX10 8DE
UK

CABI
38 Chauncey Street
Suite 1002
Boston, MA 02111
USA

Tel: +44 (0)1491 832111
Fax: +44 (0)1491 833508
E-mail: info@cabi.org
Website: www.cabi.org

Tel: +1 800 552 3083 (toll free)
Tel: +1 (0)617 395 4051
E-mail: cabi-nao@cabi.org

© CAB International 2013. All rights reserved. No part of this publication may be reproduced in any form or by any means, electronically, mechanically, by photocopying, recording or otherwise, without the prior permission of the copyright owners.

A catalogue record for this book is available from the British Library, London, UK.

Library of Congress Cataloging-in-Publication Data

Raspberries / edited by Richard C. Funt, Harvey K. Hall.
  p. cm. -- (Crop production science in horticulture series ; 23)
  Includes bibliographical references and index.
  ISBN 978-1-84593-791-1 (alk. paper)
  1. Raspberries. 2. Raspberry industry. I. Funt, Richard C. II. Hall, H. K. (Harvey K.) III. Series: Crop production science in horticulture ; 23.

SB386.R3R36 2013
664'.804711--dc23

2012030954

ISBN: 978 1 84593 791 1

Commissioning editor: Sarah Hulbert
Editorial assistant: Alex Lainsbury
Production editor: Tracy Head

Typeset by Columns Design XML Ltd, Reading, UK.
Printed and bound in the UK by CPI Group (UK) Ltd, Croydon, CR0 4YY.

Contents

Contributors		vii
Preface		ix
Dedication		xi
Acknowledgments		xiii
Conversion Factors for Weights and Measures		xv
1	**Raspberries** *Kim Hummer and Harvey K. Hall*	1
2	**Growth and Development** *Richard C. Funt*	21
3	**Climatic Requirements** *Harvey K. Hall and Tim Sobey*	33
4	**Site Selection** *Harvey K. Hall and Tim Sobey*	45
5	**Cultivar Development and Selection** *Courtney Weber*	55
6	**Nursery Production of Plants** *Alison Dolan*	73
7	**Propagation** *Courtney Weber*	83

Contents

8   Site Preparation, Soil Management and Planting   91
     Marvin Pritts and Eric Hanson

9   Soil and Water Management   103
     Richard C. Funt and David S. Ross

10   Pruning and Training   121
     Richard C. Funt

11   Pest and Disease Management   133
     Richard C. Funt

12   Crop Production   157
     Michele Stanton

13   Postharvest Physiology and Storage of Raspberries   177
     Penelope Perkins-Veazie

14   Marketing of Raspberries   191
     Gail Nonnecke, Michael Duffy and Richard C. Funt

15   Raspberry Farm Management and Economics   201
     Richard C. Funt

16   World Raspberry Production and Marketing: Industry Changes and Trends from 1960 to 2010   213
     Chaim Kempler and Harvey K. Hall

Appendix 1: Windbreaks   235
     Michele Stanton

Appendix 2: Fertigation Systems, Calibration, Drip Irrigation and Fertigation   245
     David S. Ross and Richard C. Funt

Glossary 1: Biological Terms   263

Glossary 2: Business Terms   269

Index   273

Color plates for Chapters 1–8 can be found following p. 54.
Color plates for Chapters 10–16 can be found following p. 190.

CONTRIBUTORS

Alison Dolan, PhD, Plant Pathologist, The James Hutton Institute, Invergowrie, Dundee DD2 5DA, UK; Alison.Dolan@hutton.ac.uk

Michael Duffy, PhD, Iowa State University, Department of Economics, Ames, IA 50011, USA; mduffy@iastate.edu

Richard C. Funt, PhD, Professor Emeritus, Department of Horticulture and Crop Science, The Ohio State University, Columbus, OH 43210, USA; richardfunt@sbcglobal.net

Harvey K. Hall, MS, Shekinah Berries Ltd, 1 Clay Street, Motueka 7120, New Zealand; hkhall@clear.net.nz

Eric Hanson, PhD, Michigan State University, 338 PSSB Department of Horticulture, East Lansing, MI 48824, USA; hansone@msu.edu

Kim Hummer, PhD, Research Leader, USDA ARS National Clonal Germplasm Repository, 33447 Peoria Road, Corvallis, OR 97333-2521, USA; Kim.Hummer@ars.usda.gov

Chaim Kempler, MSc, Agriculture and Agri-Food Canada, 6947 Highway 7, PO Box 1000, Agassiz, BC V0M 1A0, Canada; chaim.kempler@agr.gc.ca

Gail Nonnecke, PhD, Iowa State University, Department of Horticulture, Ames, IA 50011, USA; nonnecke@iastate.edu

Penelope Perkins-Veazie, PhD, Plants for Human Health Institute, Department of Horticulture Science, North Carolina State University, 600 Laureate Way, Suite 1321, Kannapolis, NC 28081, USA; Penelope_perkins@ncsu.edu

Marvin Pritts, PhD, Department of Horticulture, Cornell University, Ithaca, NY 14853, USA; mpp3@cornell.edu

David S. Ross, PhD, Professor Emeritus, Department of Environmental Science and Technology, University of Maryland, College Park, MD 20742, USA; dsross@umd.edu

Tim Sobey, TSA Consultants, Redbank, Herefordshire, Ledbury HR8 2JL, UK; tsobey@ledburybband.co.uk

Michele Stanton, MS, Instructor of Horticulture, The University of Cincinnati, College of Design, Art, Architecture and Planning, Cincinnati, OH 45221, USA; michele.stanton@uc.edu

Courtney Weber, PhD, Small Fruits Breeder, Department of Horticulture, Cornell University, New York State Agricultural Experiment Station, 630 W. North Street, Geneva, NY 14456, USA; caw34@cornell.edu

PREFACE

Raspberries are found on every continent except Antarctica. The first raspberries were cultivated about 450 years ago, making raspberries a relatively new crop as compared to other fruits, such as apples or peaches. At most, two cultivars were grown in England 350 years ago. Cultivars were unnamed until after 1800. During the late 19th century and later in the 20th century, raspberry plants were taken from the 'wild' and from gardens and crossed to create offspring with large fruit size and disease resistance.

Commercially cultivated raspberries are derived from the European red raspberry (*Rubus idaeus* L.) and the North American red (*Rubus strigosus* Michx.) and black (*Rubus occidentalis* L.) raspberries. While *R. idaeus* L. is native to Europe, it is a close relative to those in Japan and northern Asia. The American red raspberry (*R. strigosus* Michx.) is native to the mountain areas along the east coast (Georgia to Pennsylvania to Maine), from northern central states (the Dakotas and to the far north in Canada) and west to British Columbia, Canada. It has greater winter hardiness than the European species. Because they are so similar and hybridize easily, some taxonomists consider them to be the same species. The American black raspberry has a more southern range than the wild American red raspberry. Hybridization can occur between the black and red species, producing offspring with purple fruit. Mutations frequently occur in the red raspberry, giving rise to yellow or orange fruit. The extreme diverse nature of the genus *Rubus*, with hundreds of species of raspberry alone, has provided breeders the opportunity to create new kinds of cultivars by crossing among distantly related species. In US breeding programs, the first crosses were made at various individual state experiment stations in 1909 and at the US Department of Agriculture (USDA) at Beltsville, Maryland, Corvallis, Oregon and Cheyenne, Wyoming. Crosses were also made in Canada as early as 1893 and in England and Canada by private breeders.

Breeders during the 20th century have developed new cultivars to crop over a longer ripening season, and to resist or tolerate diseases of the fruit, shoot or root. This greatly improved yield and enhanced fruit quality, both for fresh market and processing. Fruit specialists in the USA ran long-term cultivar trials to demonstrate the new plants and how they adapted to new climatic locations. With this local and regional information, growers did not have to go through trial and error themselves in the selection of more productive or more marketable fruit. As the raspberry industry grew and was marketed to off-farm consumers, the economic impact was substantial. With the expansion of the Land Grant system in the USA in the 1930s, plant physiologists, plant pathologists, entomologists and virologists investigated and enhanced the production of virus-indexed and essentially disease-free (clean) nursery plants. More recently the production of plants from tissue culture allows for a less expensive system of propagation for the commercial grower.

It is important for the gardener, grower, student and scientist to understand the growth and development of a raspberry plant to ensure a healthy and productive planting for home garden or commercial use. A key starting point is the purchase of disease-free (clean) plants. The planting of healthy stock is essential for high and early yields for the home garden and for a profitable commercial business operation. Quality plants will be productive when grown in a good environment with a well-drained soil or growing medium, a high level of minerals and organic matter, ample high-quality water, and weed-free fields.

The authors in this book have used their talents to explain the many complex steps involved with producing a crop of raspberries that is fresh and wholesome. Further, the authors have had considerable experience in writing and speaking to growers on the best management practices under a wide range of different growers' climatic conditions. Much of the research data gleaned from hundreds of bulletins, conference proceedings, newsletters and reports during the 20th century is presented. References are presented at the end of each chapter to allow a deeper understanding of the topics presented here. An understanding of the biology, economics, technology and consumer preference of raspberries is the major objective of this publication.

Richard C. Funt
Harvey K. Hall
April 2012

DEDICATION

This book is dedicated to the amazing collection of researchers, growers and marketers, who during the 19th and 20th centuries brought the raspberry to millions of people around the world, and to those present and future generations who continue to improve the raspberry plant, create cost-effective production systems to improve the harvest and marketing of the raspberry, and continually learn how the raspberry can improve human health in the 21st century.

Specifically, we are indebted to plant breeders, plant pathologists, virologists, entomologists, plant scientists, agricultural economists, agricultural engineers, researchers, outreach specialists, agricultural editors and to many others who have spent lifetimes working to improve raspberry cultivars and grow the raspberry industry worldwide. Passionate people have produced nursery plants, conducted field trials, tested and sought chemical registration for use on this minor fruit crop. Many graduate students have contributed field and medical research projects and published their results in scientific publications and bulletins.

Agricultural engineers improved drip irrigation by using ultraviolet-resistant material, created uniform distribution of water over long distances and hilly terrain, and developed daily watering schedules to optimize yields. They improved the effectiveness and efficiency of pesticide application equipment as well as raspberry mechanical harvesters that can reduce the need for hand harvesting, and they have designed and built polythene tunnels for year-round production and sustained fruit quality, in spite of the vagaries of weather and climate variations.

Land Grant small fruit specialists tested insecticides, fungicides and herbicides that could be safely used on raspberry plants and worked closely with the US Environmental Protection Agency (USEPA) for crop clearance of new complex chemicals and to provide safe working conditions for growers and their employees. The Land Grant system worked hand in hand with producers of quality fruit to supply a safe and affordable product in the marketplace for consumers to enjoy. And, just as important are the scientists who wrote vast amounts of new information in an understandable form for all to understand and the communications staff at the Land Grant system and USDA who edited and formatted the information for growers and the agricultural industry.

The authors of this book are especially indebted to the collaborative research and outreach from the laboratory testing and human clinical trials being conducted by medical biochemists, particularly over the past 15 years. In this collaboration, there are horticulturists, food scientists and human nutritionists discovering that a diet containing raspberries can impact the human immune system and prevent certain cancers. This will have a great impact on human health in the 21st century.

We would be remiss if we did not include the many growers, nursery plant producers and equipment company representatives who walked the fields to solve problems, organized grower groups and established winter grower meetings and university conferences, and those who sought government funding to support virus-free and disease-free programs, raspberry research, and the medical investigations and trials that benefitted the general public. We have had the distinct pleasure of meeting and working alongside many of the people that have been mentioned in this dedication over the past 60 years. Along with them we dedicate this book to our parents and grandparents who were our first teachers, who taught us the many practical commercial field applications and the economics and marketing of raspberries.

It is now time to collectively bring together the vast amount of information from the laboratory, from applied field studies, from the medical clinical research, and from the postharvesting handling and marketing of a perishable commodity for the enjoyment and healthful benefits for millions of people living on this planet. Therefore, the purpose of this book is to cover all aspects of the raspberry for students, growers, berry industry suppliers, and breeders of the 21st century who will continue the effort to learn, produce quality fruit, and provide agricultural supplies and to create a better, larger berry with improved flavor and increased levels of medicinal components.

Richard C. Funt
Harvey K. Hall
April 2012

Acknowledgments

Funding for the color plates was kindly provided by the following companies:

North American Raspberry & Blackberry Association (NARBA)
www.raspberryblackberry.com

GreenStar Cooperative, Inc.
www.greenstarcoop.net

Indiana Berry and Plant Company, LLC
www.indianaberry.com

Kathleen Demchak, The Pennsylvania State University, and Shirley M. Funt, The Ohio State University, are gratefully acknowledged for editing and reviewing of chapters.

Disclaimer

The publishers, editors and authors have attempted to provide details of the latest production methods, growing techniques, pest and disease control measures and fertilizer management with raspberries, and will not be held liable for use or misuse of information provided herein. The reader or user is encouraged to consult a local professional advisor on the suitability of application of products or management techniques in his or her region. Where use of named products or chemicals is described, no discrimination towards other products is intended or implied.

Conversion Factors for Weights and Measures

	SI units	US units
Length	1 millimeter (mm)	0.039 inch (in)
	1 centimeter (cm) (10 mm)	0.39 inch (in)
	1 meter (m) (100 cm)	39.4 inches (in)
	1 kilometer (km) (1000 m)	0.62 mile (mi)
Area	1 square centimeter (cm^2)	0.155 square inch (in^2)
	1 square meter (m^2)	1.2 square yards (yd^2)
	1 hectare (ha) (10,000 m^2)	2.471 acres (ac)
	1 square kilometer (km^2) (100 ha)	247.1 acres (0.386102 mi^2)
Weight	1 gram (g)	0.035274 ounces (oz)
	1 kilogram (kg) (1000 g)	2.204623 pounds (lb)
	1 tonne (metric tonne, T) (1000 kg)	1.10231 tons (US)
Volume	1 milliliter (ml)	0.033814 fluid ounces (fl oz)
	1 liter (l) (1000 ml)	1.056688 quarts (qt)
	1 cubic meter (m^3) (1000 l)	264.172052 gallons (gal) (US)
	1 megaliter (Ml)	0.810708 acre feet (ac*ft)
Ratios	1 gram per m^2 (g/m^2)	0.003277 ounce per ft^2
	1 gram per m^3 (g/m^3)	0.0009988 ounce per ft^3
	1 tonne per hectare (t/ha)	0.446090 tons per acre (t/ac)
	1 tonne per hectare (t/ha)	892.179122 pounds per acre (lb/ac)
	1 gram per liter (g/l)	0.008345 pounds per gallon (lb/gal)
	1 gram per liter (g/l)	1 part per million (ppm)
Energy	1 joule (J)	0.0009478 btu
	1 kilojoule (kJ)	0.9478 btu
Temperature	°C (Celsius)	(°F − 32) ÷ 1.8

Conversion Factors

	US units	SI units
Length	1 inch (in)	2.54 centimeters (cm)
	1 foot (ft) (12 in)	30.48 cm
	1 yard (yd) (3 feet)	0.9144 meters (m)
	1 mile (mi) (5280 feet)	1.609344 kilometers (km)
Area	1 square inch (in^2)	6.5 square centimeters (cm^2)
	1 square foot (ft^2) (144 in^2)	930 (cm^2)
	1 square yard (yd^2) (9 ft^2)	0.84 square meters (m^2)
	1 acre (ac) (43,560 ft^2)	0.405 hectares (ha)
	1 square mile (640 acres)	259 ha (2.589988 km^2)
Weight	1 ounce (oz)	28.3 grams (g)
	1 pound (lb) (16 oz)	0.454 kilograms (kg)
	1 ton (US) (t) (2000 lb)	0.907 metric tonnes (T)
Volume	1 tablespoon (tbsp) (3 teaspoon (tsp))	14.79 milliliters (ml)
	1 fluid ounce (fl oz) (2 tbsp)	29.57 ml
	1 cup (c) (8 fl oz)	0.237 liters (l)
	1 pint (pt) (2 cups)	0.473 l
	1 quart (qt) (2 pts, 4 cups)	0.946 l
	1 gallon (US) (gal) (4 qt)	3.8 l
	1 cubic foot (cu ft) (7.45 gal)	28.3 l
Ratios	1 ton/acre (t/ac)	2.241702 t/ha
	1 pound/acre (lb/ac)	0.001121 t/ha
	1 pound/gallon (lb/gal)	119.826427 g/l
	1 ounce/ft^2	305.1517 g/m^2
	1 ounce/ft^3	1001.1539 g/m^3
Energy	1 btu	1055.05585 joules (J)
	1 btu	1.05505585 kilojoules (kJ)
Temperature	°F (Fahrenheit)	$(1.8 \times °C) + 32°$

Weights and measures for raspberry fruit

	US units	SI units
Volume of fruit	½ pint (6 oz)	170.5 g
	1 pint (12 oz)	341 g
	1 quart (24 oz)	681.6 g
Packaging of fruit	1 master (12 pints or 6 quarts in two layers)	4.0896 kg
	1 flat (12 pints or 12 ½ pints in one layer)	4.0896 kg or 2.0448 kg

	SI units	US units (volume of fruit)
Weight of fruit	125 g	4.4 ounces (oz)
	250 g	8.8 oz
	500 g	1 lb 1.6 oz
	1 kilogram (kg) (1000 g)	2 lb 3.2 oz

Abbreviations

SI units: mm, millimeter; cm, centimeter; m, meter; km, kilometer; ha, hectare; mg, milligram; g, gram; kg, kilogram; ml, milliliter; l, liter.

US units: in, inch; ft, foot; yd, yard; mi, mile; ac, acre; oz, ounce; lb, pound; t, ton; T, metric tonne; tbsp, tablespoon; tsp, teaspoon; c, cup; pt, pint; qt, quart; gal, gallon.

RASPBERRIES

KIM HUMMER[1]* AND HARVEY K. HALL[2]

[1]*USDA ARS National Clonal Germplasm Repository, Corvallis, Oregon, USA;* [2]*Shekinah Berries Ltd, Motueka, New Zealand*

INTRODUCTION

Recorded history tells us that humans have been interacting with thorny rambling and upright plants, such as raspberries and their relatives, over the course of several millennia. These plucky plants were described by the ancient Greeks and Romans, including Aeschylus, Hippocrates, Cato the Elder, Ovid and Dioscorides.

Raspberry seeds dating back to Roman times have been discovered in ancient forts in Britain. Roman soldiers probably spread the cultivation of red raspberries throughout Europe as they marched. In Russia, raspberries were consumed in combination with European blueberries before Moscow was established in 1147 (Zhukovskii, 1950). Red raspberries were popularized and selected in Europe throughout the Middle Ages. Cultivation of the red raspberry began about this time in Europe. By the time of the ancient Greeks, thornless forms were known and described (Hummer and Janick, 2007; Hummer, 2010).

In old English terms, raspberries were known as 'hyndberries', 'raspis' or 'raspes', giving rise to the modern term raspberry. Yellow-fruited forms, a color mutant of the common red-fruited raspberry, were being documented in Europe by the late 1500s. John Parkinson, an English herbalist–botanist of the period, described red, white and thornless 'raspis-berries' suitable for the English climate (Jennings, 1988). Raspberries were grown for the London market by gardeners of Cheswick and Bretford in the time of Shakespeare (Jennings, 1988). Richard Weston, an English botanist, mentioned a 'twice-bearing kind' in 1780 (Jennings, 1988).

Both Europeans and Asians recognized the medicinal value of raspberries. The Asian raspberry, specifically *Rubus chingii* Hu, *fu-pen-tzu*, was documented in the *Ming Yi Bie Lu*, a Chinese herbal dating to the end of the Liang Dynasty

*Kim.Hummer@ars.usda.gov

before 300 CE (Hsu et al., 1986). Another Asian raspberry, R. parvifolius L., tzu-po, is a traditional folk medicine of Taiwan (Hsu et al., 1986). The Indian medicinal uses of Rubus include application of the astringent diuretic action of the leaves and bark. Raspberry leaves are combined with many other herbs to produce 'kasaya'.

The native Rubus species of Europe, such as the blackberry, R. fruticosus, and the red raspberry, R. idaeus, are not endemic in India but were brought there through human exchange from east to west along the Silk Route. Many web-related nutraceutical applications of Ayurvedic medicine promote the use of accessible European Rubus species, such as the red raspberries, R. idaeus L., rather than Asian ones, e.g. R. coreanus, R. crataegifolius or R. parvifolius. Raspberries were exported to the American colonies in the 1700s. Probably one or more selections of the American red raspberry were cultivated in the USA by 1800 (Darrow, 1937). In 1771, W. Prince was the first to sell raspberry plants commercially in New York; his son W.R. Prince published a pomological manual in 1832 with descriptions of 20 cultivars (Jennings, 1988).

In 1853, the American Pomological Society recommended four European red raspberries for fruit cultivation, and by 1891, recommended 14 European and six American cultivars (Hedrick, 1925). The European raspberries proved less adaptable to the extremes of American summer and winter weather than did the local forms.

Improvements in red raspberries occurred when European red raspberries were crossed with American red raspberries (Jennings, 1988). Botanists have differed over whether or not American and European raspberries differ at the order of species or subspecies level. American native red raspberries differ from European ones by having hardier canes, which are thinner, taller and more erect. Fruit of American raspberries are usually round and seldom conical, and sometimes fruit in the wild are larger than that of the European subspecies (Jennings, 1988). The cultivar 'Cuthbert', most likely a cross between R. idaeus 'Hudson River Antwerp' and a native R. strigosus, was discovered about 1865. 'Cuthbert' remains a valuable cultivar and source of useful traits for present-day breeders. The present botanical terms recommend that these differences are at the species level (USDA, 2011).

CLASSIFICATION AND DISTRIBUTION

Raspberries are included within the genus Rubus L., under the family Rosaceae Juss. The plants within Rubus are diverse. Because of this, botanists have subdivided the genus subgenera and sections. By morphological definition, raspberry fruits consist of clusters, or aggregates, of small drupe fruits called drupelets. The aggregate of drupelets are held together by interlocking hairs and the whole group of them separate as a unit (called an 'aggregate fruit') from a conical receptacle when the ripe fruit is picked.

Rubus is placed in the *Rosaceae* tribe *Rubeae*. Five *Rubus* subgenera (*Idaeobatus, Cylactus, Anoplobatus, Chamaemorus* and *Malachobatus*) have the raspberry-type fruit character, i.e. the aggregate fruit separates from the receptacle. Many raspberry-type species have contributed to the development of cultivated raspberry fruits (Table 1.1).

Until the 21st century, the main economic development in the commercial raspberry had taken place in subgenus *Idaeobatus*. The additional subgenera are becoming larger parental contributors for the raspberry gene pool as breeding efforts are put into searching for useful and unusual traits to add to the raspberry.

Idaeobatus contains about 200 wild species. Most of these raspberry species are diploid, i.e. have two sets of chromosomes, with a count of $2n = 2x = 14$, but a few triploids (three sets of chromosomes) and tetraploids (four sets) exist (Thompson, 1995a,b, 1997). *Idaeobatus* species are found in northern Asia, but are also located in Africa, Australia, Europe and North America (Jennings, 1988). The greatest diversity is found in south-west China, the likely center of origin of the subgenus.

Raspberry species from the *Rubus* subgenera *Cylactis* (Arctic raspberry), *Idaeobatus* (raspberries), *Anaplobatus* (flowering raspberries) and *Orobatus* are native throughout the northern hemisphere, primarily in the temperate, cool temperate, cold temperate and sub-Arctic climatic zones. A few species are also found in the tropics and the southern hemisphere. The main center of diversity is in Asia, principally within the borders of modern-day China, but extending southward through the Himalayas, into Bhutan and India and east into Korea and Japan. Locations more distant from this region have fewer species and less variability within the localized species. In contrast, some distant species, for example the flowering raspberries and black raspberries of North America, show many differences from species found near the center of diversity. *Malachobatus*, an unusual subgenus of *Rubus* with entire lobed leaves and large bracts, is also found in China, though some species are distributed in Malaysia and Australia.

Raspberries, like other *Rubus* species, are adapted to consumption and spread by animals, particularly by birds and small mammals, which provide effective seed treatment. Normally seeds require scarification and stratification for germination to occur. Seed germination is promoted by light. Stratification requirements may be reduced in species adapted to either very cold or very warm conditions, allowing immediate germination when environmental conditions are suitable. The spread of raspberry species was probably from the Chinese center of raspberry diversity to the entire northern hemisphere, and speciation has occurred in response to isolation combined with genetic drift, selection pressure from different environments and segregation among seedling populations, spread and propagated by birds.

Table 1.1. *Rubus* species cited as useful in raspberry breeding programs throughout the world (modified from Daubeny (1996) and Finn (2002), with additions).

Species	Traits of interest
R. arcticus L. (and *R. stellatus*)	Earliness in primocane fruiting, good flavor, aroma, winter hardiness
R. biflorus Buch.-Ham. ex Sm.	Low chilling requirement, resistance to drought, tolerance to high temperature, leaf spot resistance, cane spot resistance
R. chamaemorus L.	High ascorbic acid content, winter hardiness
R. chingii Hu	Chinese traditional medicine
R. cockburnianus Hemsley	Very high numbers of fruits per lateral, late ripening
R. corchorifolius L.	Early ripening, erect habit
R. coreanus Miq.	Productivity, vigor, *Amphorophora idaei* and cane disease resistance, excellent black forms for crosses with black raspberry and yellow apricot forms for crosses with red raspberry
R. crataegifolius Bunge	Bright red fruit, easy plugging, pest and disease resistance, suitability for machine harvesting, early and condensed ripening
R. flosculosus Focke	Vigor, virus resistance, color variations
R. geoides	Low-growing ornamental
R. idaeus (including *R. strigosus*)	Late fruiting season, resistance to root rot, vigor, winter hardiness, flavor, resistance to the resistance-breaking strain of aphids, resistance to cane diseases, resistance to spider mites, ellagic acid content
R. innominatus Moore (including var. *kuntzeanus*)	Productivity, high fruit number/lateral, low chilling requirement, resistance to drought, high temperature, leaf spot, cane spot, cane beetle
R. lasiocarpus Focke	Disease resistance
R. lasiostylus Focke	Large fruit size, ease of harvest, fruit cohesiveness
R. leucodermis Doug ex Torrey & A. Gray	Vigor, disease tolerance, particularly viruses and *Verticillium*
R. mesogaeus	Resistance to cane Botrytis, cane blight, cane midge
R. multibracteatus H. Lev. & Vaniot	Upright growth, bright red fruit
R. niveus Thunb.	Resistance to fruit rot, fruit quality, low chilling, fruit firmness, primocane fruiting
R. occidentalis L.	*Phytophthora* resistance, resistance to biotypes 1–4 of *A. idaei*, firmness, easy plugging, late ripening
R. odoratus L.	Earliness in primocane fruiting, ornamental potential, leaf and stem spot resistance, aphid resistance
R. parviflorus Nutt.	Resistance to root rot and aphids
R. parvifolius L.	Vigor, productivity, primocane fruiting
R. phoenicolasius Maxim.	Possible disease resistance, resistance to cane beetle, powdery mildew, root rot

Species	Traits of interest
R. pileatus Focke	Resistance to cane blight, cane midge, cane Botrytis, spur blight, fruit rot, root rot, fruit flavor
R. pungens Cambess.	Possible disease resistance, early ripening floricane fruit, winter hardiness, resistance to spur blight
R. spectabilis Pursh.	Earliness in summer and primocane types, bright red fruit, easy plugging, *Phytophthora*, cane Botrytis and *Didymella applanata* field resistance, resistance to aphids
R. sumatranus Miq.	High drupelet numbers

Red raspberries

Rubus idaeus L. is the scientific name that Linnaeus gave to the typical European red raspberry. This species originated near the Ida Mountains of Asia Minor, now Turkey (Hedrick, 1925; Jennings, 1988; Roach, 1988; Daubeny, 1996).

While a range of raspberry species have been cultivated in different parts of the world, *R. idaeus* L. is the main domesticated type. Modern-day cultivars are derived mainly from the European red raspberry (*R. idaeus* subsp. *vulgatus*) and the North American red raspberry (*R. idaeus* subsp. *strigosus*, also known as *R. strigosus*).

Cultivated *R. idaeus* differs significantly from original wild types by having larger, sweeter fruit. However, the entire improved raspberry germplasm pool is based on a limited number of accessions of *R. idaeus* and *R. strigosus*. This was examined in detail by Dale *et al.* (1993), who determined the diversity of 137 cultivars of known pedigrees released between 1960 and 1993. These were derived from 50 founding clones, of which half are represented only once or twice among the 137 descendants. In the remainder, two species were used for crossing and may have made no genetic contribution. Also, there were three pairs of parents and offspring, reducing the total number of founding clones for the majority of raspberry cultivars to just 20. Among these 20 clones, 'Hudson River Antwerp' was in the background of 110 of the cultivars, 'Lloyd George' of 108, and 'English Globe' and 'Highland Hardy' of 90 (through 'Marlboro').

A raspberry pedigree database with almost 6000 entries was examined to determine how many of the cultivars listed in Hedrick's *The Small Fruits of New York* (1925) were recorded as being part of the pedigree of cultivars developed since that time. This database contains pedigrees of most of the world's raspberry cultivars, plus significant numbers of selections from several major international raspberry breeding programs. The total number of cultivars from *The Small Fruits of New York* represented in the database and used in breeding was 25, of which 12 were cultivars developed by private or public breeders and the remainder represented unique accessions of either *R. idaeus* or *R. strigosus*. The 25 cultivars were used 204 times in breeding and among them 'Lloyd George' was used more than 25% of the time; 'Latham', 'Cuthbert' and 'Newman' combined for a further 35% and 'Viking', 'Ranere', 'June' and 'Herbert' an additional 15%. In total, over 75% of the genetic input for modern cultivars from this material was derived from only eight cultivars. The remaining almost 25% brought the total of founding parents up to, at most, 25 accessions from the wild incorporated prior to 1925. These were 12 accessions of *R. idaeus*, 11 of *R. strigosus*, two of *R. occidentalis*, one of *R. innominatus* and the remaining accession an unknown species from China.

In total, more than 400 clones were identified in Hedrick's 1925 review. Many of these were generated through the actions of man, either by deliberate crossing or through chance seedlings associated with plantings of improved germplasm from either North America or Europe. Eighty-three of these were listed with at least one known or assumed parent. Nevertheless, almost all of this diversity has been lost to modern-day breeders.

Since 1925, the larger raspberry breeding programs in Europe and North America have frequently introduced new germplasm from *R. idaeus* and from *R. strigosus*. A small number of new genotypes were successfully used in breeding, especially in the European breeding programs, in spite of many populations being grown for assessment (Rozonova, 1939; Haskell, 1960; Jennings, 1963; Rousi, 1965; Oydvin, 1970; Keep, 1972; Mišić and Tesovic, 1973). In North America similar assessments of seedling populations have been made at Prince George in British Columbia (Van Adrichem, 1972). Further efforts to introgress new genetics from *R. strigosus* have been made at the Pacific Agri-Food Research Centre of Agriculture and Agri-Food Canada (AAFC) in British Columbia, (PARC-BC), Cornell University in New York and Oregon (ORUS) programs (Kempler and Daubeny, 2008; C. Weber, n.d., personal observation). The use of *R. strigosus* has primarily been through selection for various traits and introgression of these into material better suited for commercial production. More diversity is available for breeding use, but lack of resources limits this from being done systematically on a significant scale (Daubeny, 1996).

A number of traits have been assembled in modern-day cultivars that occur independently, though infrequently, in the wild. Traits of primary importance include self-fertility, resistance to diseases (especially viruses), resistance to insects, such as aphids which vector (transmit) viruses, large fruit size, strong upright growth and reduced numbers of new canes. Recently, primocane (first year) fruiting has become highly valued. These attributes are essential in modern raspberry cultivars, enabling the plants to grow in a manageable fashion, to produce high yields and to survive for an extended period in expanding regions of cultivation.

Red raspberries were initially brought into cultivation in North America using *R. idaeus* derived clones that were planted there after the settlement of Europeans (Jennings, 1988). Local *R. strigosus* selections proved superior in some traits, especially in climatic adaptation and resistance to disease. During the 19th century, a wide range of raspberry cultivars was developed on each side of the Atlantic using both subspecies.

Political boundaries also affected the domestication and development of the raspberry, especially in the isolation of Eastern Europe from Western Europe and North America in most of the 20th century. This has meant that only the earliest improved cultivars using germplasm from both sides of the Atlantic were available there. Most of the cultivars developed in the former Union of Soviet Socialist Republics (USSR) and allied countries have not had the benefit of access to many of the more recent cultivars developed. Thus, there have been very intensive breeding efforts in Eastern Europe, using early Western germplasm from Europe and North America, local selections and cultivars, and limited new introductions of East Malling Research (EMR) and Scottish Crop Research Institute (SCRI) germplasm which have not been sufficiently winter hardy for climatic conditions in most of the former USSR. Nevertheless, cultivars have been produced which are able to withstand local climatic conditions. New cultivars have invigorated this industry, which still produces almost half of the world's raspberries. Trials of new cultivars from the Russian Federation are under way in China, but not yet in Western countries.

Improved plant traits among domesticated red raspberries include vigorous upright growth, reduced cane number, resistance to diseases, good lateral attachment, high fruit number per lateral, adaptation to specific environmental conditions and enhancement of the primocane fruiting trait found only with weak expression in the wild. Fruiting has been improved with the selection of self-fertile types with large fruit size, improved firmness, enhanced flavor and color, high yield and ease of harvest. Modern-day plant breeding continues the process of domestication of the red raspberry, improving the cultivated types to make them more appreciated by the consumer and profitable for the grower.

Raspberry life cycle

Most raspberry species take 3 years from planting to reach maturity and their full production potential. These plants can remain productive for 8–12 years. By the end of that time, viral diseases frequently have built up within the plants and limit their lifespan. Commercial producers should replace their plantings periodically for maximum productivity.

Raspberries produce either one or two crops of fruit annually, depending on the cultivar. The summer-bearing raspberries produce a single crop of fruit on canes that arose the previous year beginning in early to mid-summer. These fruiting canes are called floricanes. Autumn-bearing (fall-bearing) or everbearing raspberries produce a crop in mid-summer on the previous year's canes and another crop in late summer or early autumn on current-year canes. Canes produced in the current year are called primocanes.

At present primocane-fruiting red raspberry types are the only non-floricane fruiting raspberries cultivated commercially, but efforts are under way in Oregon and elsewhere to produce primocane-fruiting black raspberries for future commercial production. A few raspberry species, such as *R. arcticus*, *R. odoratus* or *R. ellipticus*, are perennial.

Primocane-fruiting raspberries

The primocane fruiting trait in raspberries occurs in wild *R. idaeus*, *R. strigosus* and *R. occidentalis*, but the natural expression of this character is limited to a small amount of fruit at the cane tips, often late in the autumn. The primocane fruiting trait also occurs in cold climate herbaceous species, including *R. arcticus* L., *R. chamaemorus* L. and *R. saxatilis* L., in the warm climate species *R. illecebrosus* Focke, in biennial types and also in the perennial caned species (Keep, 1961).

Among wild raspberries expressing the primocane fruiting trait, fruit size is often small, fruiting nodes are few, branching and lateral formation is limited, and fruit number per lateral or fruiting node is low (Keep, 1988). Despite the low yield and poor quality of wild primocane fruiting types, they have been cultivated in European and North American gardens for over 200 years (Mawe and Abercrombie, 1778; McMahon, 1806).

From the earliest times of the domestication of the red raspberry, plants producing some primocane fruit were noted, and early breeders made attempts to enhance this trait. A significant advance was made when varieties with primocane fruiting expression from the European red raspberry, *R. idaeus* ('Lloyd George'), and the North American red raspberry, *R. strigosus*, were interbred (Roberts and Colby, 1960). This resulted in the cultivars 'Marcy', 'Indian Summer', 'September' and 'Zeva Herbsternte', followed in later generations by 'Fallred' and 'Heritage' (Keep, 1988).

Key to the potential success of primocane fruiting raspberries are good expression of yield components and excellent fruit quality. Lewis (1941) and Haskell (1960) published theories on simple inheritance of a gene for primocane fruiting but other studies have shown that inheritance is a complex mechanism (Keep, 1988). Keep (1961, 1988) emphasized that the inheritance of primocane fruiting in raspberries was due to the interaction of both genetic and environmental factors. The realization of yield potential of primocane fruiting raspberries depends on the interaction of additive factors for season (vigor, maturity and earliness) with the local environmental conditions (climate, day length, nutrition and water availability) (Keep, 1988). Yield potential is determined by fruit size and number, fruit number per cane being dependent on bearing surface or amount of branching and lateral formation on canes combined with cane number per plant (Keep, 1988). Significant advances in earliness and in yield potential over the cultivars derived from *R. idaeus* and *R. strigosus* have been achieved by breeders at EMR with the introgression of earliness into primocane fruiting types from *R. arcticus*, *R. odoratus* and *R. spectabilis* (Keep, 1988; Knight, 1991).

Further progress in the development of primocane fruiting raspberries has been achieved with the improvement of fruit quality by introducing desirable traits from floricane fruiting types, especially through the work of Dr Derek Jennings at the SCRI and Medway Fruits, Kent, England (Jennings, 2002). The initial cross making this sort of combination was 'Glen Moy' × 'Autumn Bliss', and large numbers of seedlings from this cross were grown in the UK, Australia, New Zealand and the USA, leading to the cultivars 'Aspiring', 'Dinkum' and 'Terri Louise', and the advanced selections A83-31-B5 from Australia and GEO-1 (parent of 'Caroline') from the USA. The combination 'Comox' × 'Autumn Bliss' also produced the high-quality cultivar 'Bogong' in Australia. Further, crosses were done between the high-quality floricane fruiting raspberries 'Glen Lyon' and the selection SCRI 8216B6 and primocane cultivars and advanced selections from the EMR program to give advanced selections and cultivars in the Medway Fruits program (Jennings, 2002).

In the USA, advanced selections and cultivars from EMR, the SCRI and Medway Fruits were crossed with advanced selections in the University of Maryland, Rutgers University and the Driscoll Strawberry Associates (DSA) programs in the USA, giving rise to further levels of breeding improvement and advances in yield and quality.

In Russia, material from EMR has been used widely in the Kokino program to generate a wide range of new cultivars with upright growth, very early production, large fruit, very high yield and extended shelf life. Cultivars released from this program include 'Gerakl', 'Atlant', 'Penguin', 'Brilliantovaya', 'Golden Domes', 'Cap of Monomakha', 'Eurasia', 'Babe Leto 2' and others. These cultivars offer an exciting range of possibilities for the development of new cultivars for the future when they become available in the West.

Black raspberries

Although there are many black-fruited raspberry species, the black raspberries of commerce, also known as 'black caps', are native to North America. Named cultivars are derived from one of two species. The eastern North American black raspberry species, *R. occidentalis* L., occurs from New Brunswick to Ontario, south to South Carolina, Georgia and Alabama, and west to the Midwestern states of Missouri, Kansas, Minnesota and North and

South Dakota. The western North American black raspberry, *R. leucodermis*, occurs in British Columbia, through Washington, Oregon and into California. Selections from the wild eastern species from locations in New York, New Jersey and Pennsylvania were first domesticated and named. Crosses from these preferred plants were bred at the New York Agricultural Experiment Station in Geneva during the last century.

As with the other minor fruit crops, domestication of the black raspberry is in its early stages. The first black raspberry cultivars were selected directly from the wild in eastern North America in the 1830s. Black raspberries are diploid. They belong to the same subgenus (*Idaeobatus*) as red raspberries and can cross with species in this subgenus. Wild black raspberries are typically found in disturbed habitats near forest or trail edges where some direct sunlight is available. They frequently inhabit swales on sharp slopes with good drainage.

The black raspberry industry in North America was large in the early 1900s and has gone through some decline over the years. Black raspberry juice has long been used not only in processed products, but also as a natural colorant/dye. Pigments from black raspberries were used for many years as the dye for the stamp used to certify meat by the US Department of Agriculture (USDA). The anthocyanin pigments and fruit chemistry of black raspberries has been researched since the 1950s. Lee and Slate (1954) examined chemical properties of black raspberries to determine which cultivars were better for freezing and which were better for processed products. However, during the past 100 years, black raspberry production in the USA has declined, probably due to disease pressures and a lack of cultivars with sufficient resistance (Dossett *et al.*, 2008). Ourecky (1975) noted that breeding progress slowed dramatically as a result of a lack of genetic variability in available germplasm. He considered the lack of genetic diversity in the black raspberry to be a major limitation. He found that distinguishing existing cultivars was difficult using morphological means, and suggested that future black raspberry breeding should include germplasm from other species. Others also recognized this (Slate and Klein, 1952; Drain, 1956; Ourecky and Slate, 1966). However, the first interspecific hybridization leading to a new cultivar was 'Earlysweet' in 1996. This cultivar has *R. leucodermis* in its background (Galletta *et al.*, 1998).

Commercial black raspberry cultivars can have severe virus problems, root rot, an arching growth habit, very spiny canes and small fruit size. Little diversity has been found in existing cultivars but further traits for improvement of black raspberries can be found in wild *R. occidentalis* L. and in *R. leucodermis* Douglas ex. Torr. and A. Gray (Dossett and Finn, 2010).

The USDA began evaluating black raspberries for a breeding program in Corvallis, Oregon, in 2003 (Finn *et al.*, 2003). Through plant exchange and collection this program has obtained a broad diversity of clones and seed lots from the full extent of the range of the eastern, and a good representation of the western, North American black raspberry (Dossett *et al.*, 2008;

Dossett and Finn, 2010). This program has selected for aphid resistance, thornlessness, and *Verticillium* and other root rot resistance, besides quality and productivity.

Oregon is a leading production region for black raspberries. This crop has been observed to have a high amount of total anthocyanins of black-fruited berries (Moyer *et al.*, 2002). Oregon growers typically see a decline in production after the second harvest season and may remove fields after only three or four seasons because of decreased profitability (Halgren *et al.*, 2007). This decline has been associated with aphid-vectored viruses, specifically black raspberry necrosis virus (BRNV) (Halgren *et al.*, 2007). Disease pressure is sufficiently high that new fields are completely infected within two to three growing seasons in Oregon.

Berry breeders in the Pacific Northwest USA and elsewhere have recognized the importance of aphid resistance in new raspberry cultivars to protect against virus infection. In Europe, at least 13 major aphid resistance genes are known, and four biotypes of the aphid have been described (Keep, 1989). Resistance is derived from a variety of sources, including red and black raspberries. Keep (1989) indicated that these aphid resistance genes have been effective in slowing the spread of aphid-vectored viruses in the red raspberry at EMR. Also, at the SCRI, virus infection rates in resistant cultivars was less than 10%, while susceptible plants became completely infected within 3 years (Jones, 1976).

In North America, the primary vector of the raspberry mosaic virus complex is the large raspberry aphid *Amphorophora agathonica* Hottes. Resistance to *A. agathonica* is conferred by three genes from red raspberry. The single dominant gene *Ag1* is derived from 'Lloyd George' (Daubeny, 1996). Genes *Ag2* and *Ag3* are tandem dominant genes derived from wild *R. strigosus* Michx. populations of eastern Canada (Daubeny and Stary, 1982). Sources of partial aphid resistance have also been identified (Daubeny, 1972; Kennedy *et al.*, 1973) but these levels are less than that conferred by *Ag1*. Only a single biotype of *A. agathonica* had been observed, however; Converse *et al.* (1971) described an *Ag1* resistance breaking strain in British Columbia. New sources of resistance to *A. agathonica* are needed to slow the spread of resistance-breaking aphid biotypes and to help maintain existing resistance.

Black raspberry cultivars have been difficult to distinguish through morphological means (Ourecky, 1975). Bassil, Dossett and Finn (n.d., personal communication) have been examining molecular markers to examine differences in cultivars. Microsatellite markers have determined that many older named cultivars have been misidentified, perhaps through mis-propagations over the years, and are no longer available in the trade. Other cultivars that have been sold in different regions under the same name have multiple molecular fingerprints (M. Dossett, n.d., personal communication). Ongoing molecular research is identifying black raspberry cultivars through fingerprinting.

Development of spineless black raspberries was begun by Derek Jennings at the SCRI in 1966 with the hybridization of black raspberries and spineless red raspberries. Since then, a program of crossing, selfing and back-crossing has been used to develop spineless black raspberry types, initially at the SCRI, and then later at DSA, HortResearch and Shekinah Berries Ltd in New Zealand. Initial populations produced purple raspberries. Eliminating the purple color and dusty appearance has taken more than six generations of breeding, and a similar number of generations of improvement has been needed to restore the typical black raspberry flavor.

Attempts are being made to develop primocane-fruiting black raspberries by Peter Tallman, a private breeder, although his first cultivar release 'Explorer' (P. Tallman, n.d., personal communication) proved a commercial failure due to lack of self-fertility. Further material from this small private program is coming that will correct the deficiencies of 'Explorer'. Further, populations segregating for primocane fruiting are also amongst the material in the Oregon program and this, too, is likely to give rise to new primocane-fruiting black raspberry types.

Further advances in the development of black raspberry genetics have also been proceeding with DSA in California.

Purple raspberries

Also known as purple-cane hybrids (Darrow, 1937; Jennings, 1988), purple raspberries are domesticated cultivars with a mixture of red (*R. idaeus*) and black (usually *R. occidentalis*) raspberry genes (Pritts and Handley, 1989). This cross was given the specific status *R. neglectus* Peck. The cross seems successful only if the *R. occidentalis* parent was used as the maternal parent and the *R. idaeus* as the paternal parent. This seems attributable to an unusual one-way incompatibility in which the pollen tubes of an *R. occidentalis* father could not progress beyond the upper third of the style of an *R. idaeus* mother. This seems to be true for dark-fruited berries being crossed with red-fruited berries within the genus.

Hedrick (1925) suggests that the plants of this hybrid group vary, sometimes resembling one parent and sometimes the other. The canes are much like those of the black raspberry: they are prickly but not bristly, have an arching habit, and root at the tip. Leaves are morphologically variable and the flowering branches glabrous. Fruits are dark red to purple. The value of this group is as a productive fruit most similar to raspberry.

Most purple raspberries are less cold hardy than most red or black raspberry cultivars, and thus are not commercially viable in climates such as those in Maine, New Hampshire, Vermont or Canada. These cultivars are more tolerant of high summer temperatures than are red raspberries. 'Lloyd George', which was an outstanding parent for breeding red raspberries, was a poor parent for breeding purple raspberries (Jennings, 1988). Purple raspberries are more fertile at the tetraploid rather than the diploid level. The tetraploids produce larger fruit and are less prone to malformation from imperfect fruit set than are diploids (Jennings, 1988). Another incentive to produce purple raspberries was to introduce the primocane-fruiting trait ('Ohio Everbearing', for example) into red raspberry (Jennings, 1988). 'Purple Autumn' was an important primocane-fruiting purple raspberry selected at Illinois (Jennings, 1988).

The purple raspberry cultivar 'Royalty' is now widely grown in northeast China, where it is highly productive, amenable to cane burying for winter and harvested at the early stages of ripeness when it can be sold as 'red raspberry'.

Yellow raspberries

Yellow-fruited forms of *R. idaeus* have been known for hundreds of years, and many have been named (Jennings, 1988). These light-fruiting genotypes arise from a single gene mutation that prevents the formation of normal dark-colored pigments in the fruit or in the plant. In this case, the normal red-fruited plants tend to be more vigorous and can dominate yellow forms to the point of exclusion, if not eliminated.

The black raspberry species *R. occidentalis* and *R. leucodermis* frequently produce sports that have white or yellow fruit (Jennings, 1988). Multiple single genes for fruit color have been described (Daubeny, 1996). Genes of yellow-fruited forms may mutate (revert) so that red-fruited canes grow.

Species native to Europe, the UK and the Middle East

The *Rubus* subgenera *Cylactis* (Arctic raspberry) is circumpolar above the Arctic Circle. Separate domestication events have taken place with other raspberry species in Europe, Asia and North America, including *R. arcticus* L., *R. crataegifolius*, *R. coreanus* and *R. occidentalis*. In Northern Europe, efforts have been made to domesticate the Arctic raspberry, *R. arcticus*, especially through the efforts of researchers at the North Savo Agricultural Experiment Station at Maaninka in Finland (Holloway, 1982). Problems encountered with this species include low productivity, small fruit size, extended fruiting season, self-infertility and strong adherence of fruit to the receptacle, making harvest difficult. Attempts have been made to cross *R. arcticus* with the red raspberry, but the hybrid retained has not been successfully back-crossed to *R. arcticus* to introgress desirable traits. Back-crosses to *R. idaeus* have taken generations to restore economic traits, by which time the distinctive flavor of *R. arcticus* was almost completely lost (Holloway, 1982).

In Sweden, attempts to produce an improved Arctic raspberry clone have utilized the Alaskan species *R. stellarcticus* in crosses with *R. arcticus* to develop improved cultivars with large flavorful fruit (Larsson, 1980). Nevertheless, these new cultivars remain self-sterile and have to be picked by cutting the calyx. There remains much to be achieved in completing the domestication of this crop to make it better suited to commercial production.

Species native to Asia and Oceania

In Asia, many raspberry types are gathered from the wild, but few attempts have been made to bring any of these species into cultivation. *R. crataegifolius* and *R. coreanus* are the exception, being cultivated with several cultivars released for growers in Korea (Hong et al., 1986; Kim et al., 2006). *R. crataegifolius* cultivars include 'Jingu Juegal' and 'Jingu Whangal'. These cultivars are not widely grown but occupy a small part of raspberry production in this region. With the black raspberry (*R. occidentalis*), a thriving industry has developed in South Korea, with over 4500 t production per year, mainly for wine making. An unusual Asian raspberry, *R. corchorifolius* (Plates 1.1a and b), demonstrates a strong upright growth habit. *R. parvifolius* (Plate 1.2) occurs commonly in China and Japan and in diploid and tetraploid forms.

R. coreanus cultivars have been developed and are being commercialized for wine making (Kim et al., 2002, 2005). *R. crataegifolius*, *R. coreanus*, *R. innominatus*, *R. niveus* and *R. pileatus* and other Asiatic species have been widely used in attempts to improve the red raspberry but only as donors for specific traits. *R. sumatranus* has an interesting deep cap-shaped fruit, although drupelets are small and dry and seeds make the fruit crunchy (Plate 1.3).

Attempts have also been made to domesticate *R. parvifolius* in the USA as a raspberry type suitable for high temperature and low chill conditions, but no commercial or even home garden cultivars have been released and no industry developed. As with the other Asiatic species, *R. parvifolius* has been used in breeding to improve the red raspberry. It has played an important part in the development of the cultivars 'Dormanred', 'Mandarin' and 'Southland', which have been adapted to conditions in south-eastern USA. It has been used to impart low chill adaptation in the HortResearch New Zealand program. *R. parvifolius* has also been valuable as a source of resistance to fluctuating winter temperatures in the BC_1 and BC_2 generations (Ballington and Fernandez, 2008).

A diverse range of Asiatic black raspberry species exists, including *R. niveus* (Plate 1.4) of China, and *R. mesogaeus* (Plate 1.5) from Japan. However, thus far they have not yet been successfully crossed with *R. occidentalis* cultivars to add diversity for black raspberry cultivar development. *R. coreanus* has been successfully used in crosses with red raspberry in the EMR program and useful traits have been introgressed into new red raspberry cultivars. Recently, *R. mesogaeus*, the Japanese black raspberry, has been collected from Hokkaido. This species may provide opportunities in the development of black-fruited raspberry selections. Jennings (1988) mentions that this species has resistance to cane Botrytis, cane blight and cane midge.

The Japanese wineberry, *R. phoenicolasius*, has become an invasive species throughout Northeastern USA. This bristly and thorny Asian native raspberry can be found growing throughout the Northeastern and Midwestern USA.

Several species of raspberries are native to the tropics, including *R. rosaefolius*, also known as the Mauritius or roseleaf raspberry, from India and Asia, and *R. ellipticus*, a small-seeded, golden raspberry species from Japan. *R. macraei* and *R. hawaiiensis* are from Hawaii. *R. hawaiiensis* can grow up to 5 m with upright, sparsely spined, perennial canes. It has pink flowers that develop into large dark purple or yellow fruit (Plate 1.6).

Species native to North America

While the major North America raspberry species, *R. strigosus* (Plate 1.7), *R. occidentalis* and *R. leucodermis*, have been discussed previously, several American raspberry species occur in the subgenus *Anoplobatus*. Although the native plants have small fruit, this species has been used in combination with European red raspberries in the development of many large-fruited cultivars.

The red (or purple) flowering raspberry, *R. odoratus*, is native to the eastern USA. This species has the good qualities of early primocane ripening, self-supporting canes, and resistance to raspberry midge and cane blight (Daubeny, 1996).

The corresponding species with western distribution in North America is the white flowering raspberry, or thimbleberry, *R. parviflorus* Nutt. This species has some compatibility difficulties with *R. idaeus*. Seedlings in the F1 population flowered profusely but failed to set fruit. Jennings and Ingram (1983) observed that the female gametes were almost completely infertile. This infertility could be overcome by colchicine treatment and working with a tetraploid-derived population. If such a population could be redeveloped, valuable genes for spinelessness, winter hardiness and cane disease resistance could be accessed (Daubeny, 1996).

The salmonberry (*R. spectabilis* Pursh) is distributed from southern Alaska, through Canada down through Northern California. This species has both early floricane and early primocane-ripening fruit. Fruit ripening is compressed and the easy to harvest fruit ripen with a bright, stable (non-darkening) red. The plants are resistant to root rot as well as the aphids, *A. agathonica*. Crosses between *R. idaeus* and *R. spectabilis* are fertile (Virdi and Eaton, 1969; Keep, 1984; Knight, 1991). Back-crosses have had mixed results (Daubeny, 1996).

SUMMARY

Cultivars that are adapted to long-cane or tunnel production have become an important part of the raspberry industry. Publicly released floricane-producing cultivars, such as 'Tulameen' in Canada and 'Glen Ample' from Scotland, though originally developed for open field production, are now being grown under tunnels. Tunnels reduce the limitation of cold wet climates with short summers, and though expensive, are the way of future production (Finn et al., 2008).

Active public raspberry breeding programs in the Pacific Northwest, New York, Canada and the UK produce, on average, about ten new cultivars each year that are being trialled, now with possible releases worldwide (Finn et al., 2008). Additional cultivars are produced by the private breeding programs for contract use in specific regions.

Retrospectives on *Rubus* breeding (Jennings, 1988; Daubeny, 1996; Finn et al., 2008) bemoaned the limited genetic resources used to develop previous industry standard raspberry cultivars during the past century. With wild collections becoming more available through international plant collections and preservation at gene banks, broader diversity is available than in previous decades. It seems that traditional breeders could now venture into the vast array of *Rubus* species relatives for potential new reservoirs of disease and pest resistance, phenological, structural and morphological improvement genes. However, just as diverse species are more available, intellectual property rights are increasingly becoming a barrier to germplasm exchange and enhancement into advanced breeding lines. Public breeding programs must now support themselves with royalties or cease existence. That means churning out the next money-making cultivar rather than enhancing the cultivated gene pool with additional genes from wild species. Private breeding programs are emerging and act on proprietary interest. The most important primocane-fruiting cultivars for 2010 belong to private companies and are proprietary, so specific cultivar names are confidential and germplasm is not shared with other programs (Finn et al., 2008).

Raspberries are a tasty nutritious fruit that are enjoyed by many. The potential for development of this fruit is great. The production regions could be expanded not only within colder climates through the use of tunnels and cold-hardy genotypes, but in warmer climates if diseases can be controlled. Breeding for improved fruit quality, plant architecture and disease/pest resistance is ongoing. To achieve this development, wider access to wild forms and a broader gene pool are essential. The accessibility to wild species through plant collection, the maintenance of wild genetic resources in international gene banks, and the enhancement of this material into breeding lines will accomplish this goal.

REFERENCES

Ballington, J.R. and Fernandez, G.E. (2008) Breeding raspberries adapted to warm humid climates with fluctuating temperatures in winter. *Acta Horticulturae* 777, 87–90.

Converse, R.H., Daubeny, H.A., Stace-Smith, R., Russell, L.N., Koch, E.J. and Wiggans, S.C. (1971) Search for biological races in *Amphorophora agathonica* Hottes on red raspberries. *Canadian Journal of Plant Science* 51, 81–85.

Dale, A., Moore, P.P., McNicol, R.J., Sjulin, T.M. and Burmistrov, L.A. (1993) Genetic diversity of red raspberry varieties throughout the world. *Journal of the American Society for Horticultural Science* 118, 119–129.

Darrow, G.M. (1937) *Blackberry and Raspberry Improvement, USDA Yearbook of Agriculture*. Government Printing Office, Washington, DC, pp. 496–533.

Daubeny, H.A. (1972) Screening red raspberry cultivars and selections for immunity to *Amphorophora agathonica* Hottes. *HortScience* 7, 265–266.

Daubeny, H.A. (1996) Brambles. In: Janick, J. and Moore, J.N. (eds) *Fruit Breeding. Volume II: Vine and Small Fruits*. Wiley & Sons, New York, pp. 109–190.

Daubeny, H.A. and Stary, D. (1982) Identification of resistance to *Amphorophora agathonica* in the native North American red raspberry. *Journal of the American Society for Horticultural Science* 107, 593–597.

Dossett, M. and Finn, C.E. (2010) Identification of resistance to the large raspberry aphid in black raspberry. *Journal of the American Society for Horticultural Science* 135, 438–444.

Dossett, M., Lee, J. and Finn, C. (2008) Inheritance of phenological, vegetative, and fruit chemistry traits in black raspberry. *Journal of the American Society for Horticultural Science* 133, 408–417.

Drain, B.D. (1956) Inheritance in black raspberry species. *Proceedings of the American Society for Horticultural Science* 68, 169–170.

Finn, C.E., Wennstrom, K., Link, J. and Ridout, J. (2003) Evaluation of *Rubus leucodermis* populations from the Pacific Northwest. *HortScience* 38, 1169–1172.

Finn, C.E., Moore, P.P. and Kempler, C. (2008) Raspberry cultivars: What's new? What's succeeding? Where are the breeding programs headed? *Acta Horticulturae* 777, 33–40.

Galletta, G.J., Maas, J.L. and Enns, J.M. (1998) Earlysweet black raspberry. *Fruit Varieties Journal* 52, 123–124.

Halgren, A., Tzanetakis, I.E. and Marin, R.R. (2007) Identification, characterization and detection of black raspberry necrosis virus. *Phytopathology* 97, 44–50.

Haskell, G.M. (1960) The raspberry wild in Britain. *Watsonia* 4, 238–255.

Hedrick, U.P. (1925) *The Small Fruits of New York*. New York State Agricultural Experiment Station, Albany, New York.

Holloway, P.S. (1982) *Rubus arcticus* L., the arctic raspberry. *Fruit Varieties Journal* 36, 84–86.

Hong, S.B., Lee, D.K., Kim, Y.H., Kong, S.J., Oh, S.D. and Kim, J.H. (1986) Characteristics of 7 Korean red raspberry lines (*Rubus crataegifolius*) selected as recommendable in Korea. Research Report of the Office of Rural Development 14, Swan, Korea, pp. 49–55.

Hummer, K. (2010) *Rubus* pharmacology: Antiquity to the present. *HortScience* 45, 1587–1591.

Hummer, K. and Janick, J. (2007) *Rubus* iconography: Antiquity to the Renaissance. *Acta Horticulturae* 759, 89–106.

Hsu, H.Y., Chen, Y.P., Shen, S.J., Hsu, C.S., Chen, C.C. and Chang, H.C. (1986) *Oriental Material Medica: A Concise Guide*. Oriental Healing Arts Institute, Long Beach, California.

Jennings, D.L. (1963) Some evidence on the genetic structure of present-day raspberry varieties and some possible implications for further breeding. *Euphytica* 12, 229–243.

Jennings, D.L. (1988) *Raspberries and Blackberries: Their Breeding, Diseases and Growth*. Academic Press, San Diego, California.

Jennings, D.L. (2002) Breeding primocane-fruiting raspberries at Medway Fruits – progress and prospects. *Acta Hort (ISHS)* 585, 85–89.

Jennings, D.L. and Ingram, R. (1983) Hybrids of *Rubus parviflorus* Nutt. with raspberry and blackberry, and the inheritance of spinelessness derived from this species. *Crop Research (Horticultural Research)* 23, 95–101.

Jones, A.T. (1976) An isolate of cucumber mosaic virus from *Rubus phoenicolasius* Maxim. *Plant Pathology* 25, 137–140.

Keep, E. (1961) Autumn-fruiting in raspberries. *Journal of Horticultural Science* 36, 174–185.

Keep, E. (1972) Variability in the wild raspberry. *New Phytologist* 71, 915–924.

Keep, E. (1984) Breeding *Rubus* and *Ribes* crops at East Malling. *Journal of Horticultural Science* 35, 54–71.

Keep, E. (1988) Primocane (autumn)-fruiting raspberries: A review with particular reference to progress in breeding. *Journal of Horticultural Science* 63, 1–18.

Keep, E. (1989) Breeding red raspberry for resistance to diseases and pests. In: Janick, J. (ed.) *Plant Breeding Reviews*, Volume 6. Wiley, Chichester, UK, pp. 245–321.

Kempler, C. and Daubeny, H.A. (2008) Red raspberry cultivars and selections from the Pacific Agri-Food Research Center. *Acta Horticulturae* 777, 71–75.

Kennedy, G.G., Schaefers, G.A. and Ourecky, D.K. (1973) Resistance in red raspberry to *Amphorophora agathonica* (Hottes) and *Aphis rubicoia* (Oestund). *HortScience* 8, 311–313.

Kim, M.J., Kim, S.H. and Lee, U. (2002) Selection of Korean black raspberry (*Rubus coreanus* Miq.) for larger fruit and high productivity. *Journal of Korean Forestry Society* 91, 96–101.

Kim, S.H., Chung, H.G., Jang, Y.S., Park, Y.K., Park, H.S. and Kim, S.C. (2005) Characteristics and screening of antioxidative activity for the fruit by *Rubus coreanus* Miq. clones. *Journal of Korean Forestry Society* 94, 11–15.

Kim, S.H., Chung, H.G. and Han, J. (2006) The superior tree breeding of *Rubus coreanus* Miq. cultivar 'Jungkeum' for high productivity in Korea. *Korean Journal of Plant Research* 19, 381–384.

Knight, V.H. (1991) Use of the salmonberry, *Rubus spectabilis* (Pursh.), in red raspberry breeding. *The Journal of Horticultural Science* 66, 575–581.

Larsson, E.G.K. (1980) *Rubus arcticus* L. subsp. × *stellarcticus*. A new arctic raspberry. *Acta Horticulturae* 112, 143–144.

Lee, F.A. and Slate, G.L. (1954) *Chemical Composition and Freezing Adaptability of Raspberries*. New York State Agricultural Experiment Station Bulletin No. 761, Geneva, New York.

Lewis, D. (1941) The relationship between polyploidy and fruiting habit in the cultivated

raspberry. *Proceedings of the 7th International Genetics Congress*, Edinburgh, 1939, p. 190.
Mawe, T. and Abercrombie, J. (1778) *Universal Gardener and Botanist: Or a General Dictionary of Gardening and Botany*. University of Cambridge, UK.
McMahon, B. (1806) *American Gardener's Calendar.* J.B. Lippincott, Philadelphia, Pennsylvania, pp. 517.
Mišić, P.D. and Tesovic, Z.V. (1973) Wild red raspberry (*Rubus idaeus* L.) in Serbia and Montenegro. *Jugosl Vocar* 7, 1–9.
Moyer, R.A., Hummer, K.E., Finn, C.E., Frei, B. and Wrolstad, R.E. (2002) Anthocyanins, phenolics, and antioxidant capacity in diverse small fruits: Vaccinium, Rubus, and Ribes. *Journal of Agricultural Food Chemistry* 50, 519–525.
Ourecky, D.K. (1975) Brambles. In: Janick, J. and Moore, J.N. (eds) *Advances in Fruit Breeding*. Purdue University Press, West Lafayette, Indiana, pp. 98–129.
Ourecky, D.K. and Slate, G.L. (1966) Hybrid vigor in *Rubus occidentalis-Rubus leucodermis* seedlings (abstract). *Proceedings of the 17th International Horticultural Congress*, 1, p. 277.
Øydvin, J. (1970) Important parent varieties in raspberry breeding. *Publication. StatensForsoksgardNjos*, pp. 1–42.
Pritts, M. and Handley, D. (1989) *Bramble Production Guide (NRAES-35)*. Northeast Regional Agricultural Engineering Service, Ithaca.
Roach, F.A. (1988) History and evolution of fruit crops. *HortScience* 23, 51–55.
Roberts, O.C. and Colby, A.S. (1960) Red and purple raspberries: their identification from plant primocanes. *University of Massachusetts Agricultural Experiment Station Bulletin* 523, 1–27.
Rousi, A. (1965) Variation among populations of *Rubus idaeus* in Finland. *Annales Agriculturae Fenniae* 4, 49–58.
Rozonova, M.A. (1939) Evolution of cultivated raspberry. *Czech Republic (Dokl.) Academy of Science URSS*, 18, 677–680.
Seely, R.J. and Martin, L. (1979) Evidence for random local spread of aphid-borne mild yellow-edge virus in strawberries. *Phytopathology* 69, 142–144.
Slate, G.L. and Klein, L.G. (1952) Black raspberry breeding. *Proceedings of the American Society for Horticultural Science* 59, 266–268.
Thompson, M.M. (1995a) Chromosome numbers of *Rubus* species at the National Clonal Germplasm Repository. *HortScience* 30, 1447–1452.
Thompson, M.M. (1995b) Chromosome numbers of *Rubus* cultivars at the National Clonal Germplasm Repository. *HortScience* 30, 1453–1456.
Thompson, M.M. (1997) Survey of chromosome numbers in *Rubus* Rosaceae: Rosoideae. *Annals of the Missouri Botanical Garden* 84, 128–163.
USDA, ARS, National Genetic Resources Program (2011) *Germplasm Resources Information Network – (GRIN)* (online database). Available at: http://www.ars-grin.gov/cgi-bin/npgs/html/tax_search.pl (accessed 5 July 2011).
Van Adrichem, M.C.J. (1972) Variation among British Columbia and Northern Alberta populations of raspberries, *Rubus idaeus* subsp. strigosus Michx. *Canadian Journal of Plant Science* 52, 1067–1072.
Virdi, B.V. and Eaton, G.W. (1969) Interspecific hybridization of the red raspberry and salmonberry. *Canadian Journal of Botany* 47, 1820.
Zhukovskii, P.M. (1950) Cultivated plants and their wild relatives: *Rubus. Commonwealth Agricultural Bureaux*, 36. State Publishing House, Soviet Science, Moscow.

Growth and Development

Richard C. Funt*

*Department of Horticulture and Crop Science,
The Ohio State University, USA*

INTRODUCTION

Raspberries are perennial plants with biennial cane (stem) growth and fruiting habit. Plants are long lived; many commercial and garden plantings can produce fruit for 10–20 years, depending on location and soil characteristics. Raspberries are different from other berries in that the fruit separate from the receptacle when mature. The leading commercial raspberry types are red fruited (*Rubus idaeus* L.), black fruited (*Rubus occidentalis*) and purple fruited (derivatives of both red and black types). Yellow and apricot-colored types (pigment-deficient selections from each of the other types) are also increasing for niche market production (refer to Plates 5.4a and b).

Breeding programs in the 20th century have enhanced the primocane fruiting trait in red raspberries. Raspberry cultivars have also been developed for winter hardiness in cold climates and for warm climates and tropical conditions. All raspberries produce fruit on floricanes that take 2 years to grow; however, primocane-fruiting red raspberries produce a large number of flowers and fruit on first-year canes. Therefore, red raspberries have two types of production systems: floricane (spring or summer) fruiting and primocane fruiting (also called autumn fruiting, everbearing, fall-bearing or remontant types).

MORPHOLOGY

Roots

The red raspberry root system develops in the upper 25–50 cm (10–20 in) of soil, with 70% of the roots in the upper 25–30 cm (10–12 in) (Fig. 2.1). Generally, the raspberry plant uses water from the upper 60 cm (24 in) of the

---

* richardfunt@sbcglobal.net

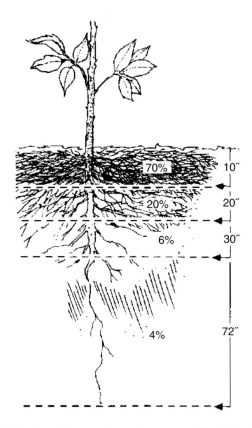

Fig. 2.1. Root distribution of the raspberry. (Courtesy of The Ohio State University.)

soil. Therefore, as the grower makes decisions about supplemental watering (irrigation), it is important to keep the root zone moist, as well as the depth of the soil (reservoir), where water can be used by the plant.

In the northern hemisphere, nursery plants are set in April. The leader bud produces vigorous canes until cold weather limits further growth; these canes then become floricanes in the second year on summer-bearing (also called June-bearing in the northern hemisphere) cultivars. Black raspberries and certain purple raspberries, which are floricane fruiting, do not spread by suckers from roots but develop buds from the crown (perennial part of the raspberry) (Fig. 2.1). Thus, when these plants do not grow in the planting year, a grower is wise to replant a black raspberry plant to optimize yields. If a red raspberry plant is missing, the planting will increase the number of plants from root buds that spread away from the crown, filling in the row where previous have died, eliminating the need for replanting in all but the most dire cases.

Canes

Raspberry canes can grow to a height of 1–5 m (3.0–16.5 ft) as a primocane, depending on soil type, cultivar and climate. These primocanes develop flower (fruit) buds at the end of the first year and produce fruit the second year. Most fruit buds for floricane cultivars are between 30 cm (12 in) and 1.5 m (60 in) from the ground. However, other cultivars may bear up to 2 m (80 in) from the ground due to cane vigor. Greater cane vigor is expressed by increased cane diameter and length of floricanes and usually greater production.

Red raspberries seldom branch in the first year unless the terminal bud is damaged or pruned. However, during long, hot summers, some red cultivars may break bud at the tip and produce side branches, which will eventually die and not produce fruit. Black raspberries and some purple raspberries produce side branches when they are primocanes (Fig. 2.2). Primocanes are always growing among floricanes after the first year (year of planting). In the second and remaining years, side branches can be forced by cutting the growing cane about 1.5–2.5 cm (0.5–1 in) below the tip as the canes grow to about 1.0 m (3 ft). Primocane fruiting red raspberries generally have canes 1.3–1.6 m (4–5 ft) tall or, in locations with longer growing seasons, 2–5 m (6–15 ft)

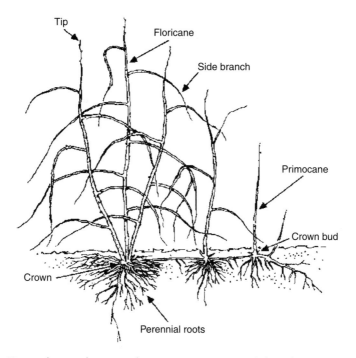

Fig. 2.2. First and second year raspberry cane. (Courtesy of The Ohio State University.)

tall, depending on the cultivar and the growth habit and fruiting season of the cultivar. Increase in plant height is arrested as flower buds are initiated in approximately 40 to 50 nodes in late spring and early summer. Plant height is also affected by warm daytime temperatures and high amounts of nitrogen, with taller plants usually having longer internodes.

Fruits

Fruiting buds are formed along upright canes of the black and floricane fruiting red raspberries, as shown in Fig. 2.3. Fruiting buds will also form on black raspberry side branches if pruned in early summer as described above. These fruiting buds remain dormant until the following year. Also, primocane fruiting red raspberries form fruiting buds (as described above). The buds do not remain dormant, with flowers and fruits developing from the top 10–15 buds in the autumn.

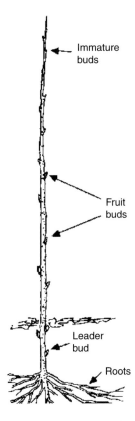

Fig. 2.3. Roots and buds of raspberry. (Courtesy of The Ohio State University.)

Raspberry flowers have five green sepals and five white petals. Many male (pollen-producing) stamens encircle a central white receptacle that contains pistils or female fruit-producing structures. Once the flower is pollinated an aggregate fruit is formed from many drupelets. Each drupelet contains a single seed. The fruit remains intact and firm during harvest (Fig. 2.4). In this text, raspberries are referred to as 'berries', but really they are not berries. Scientifically, berries have many seeds lying in a group within a single contiguous, often globe-shaped epidermis. Therefore, grapes and blueberries are berries. Raspberries have drupelets (30–150 drupelets per berry), which are arranged spirally around a receptacle (torus), and so are called aggregate fruits. When mature the drupelets adhere to each other and the entire aggregate fruit is pulled from the receptacle as a single unit with a hollow center and thimble shape.

PHYSIOLOGY

In the last half of the 20th century, many advances have been made worldwide in crossing different types of raspberries to have more productive, pest- and disease-resistant plants. More recently, breeders have been making crosses for large-sized fruit suitable for hand or mechanical harvesting (Plate 2.1). Thus, an understanding of plant growth and development, fruit bud initiation, acclimation for winter dormancy and fruit ripening is vital to a successful endeavor with raspberries. Preparing the soil and selecting a support system, soil/water management, fertilization, harvesting and marketing will be explained in other chapters in this text. An understanding

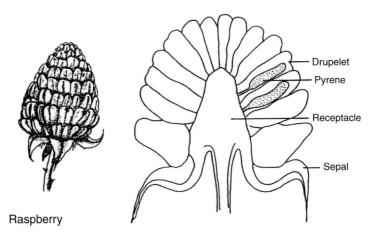

Fig. 2.4. Internal side view of a raspberry fruit. (Source: D. Jackson, New Zealand.)

of the physiology of the raspberry plant, which is discussed in these subsequent chapters, will help the reader understand the steps for maximum performance.

GROWTH AND DEVELOPMENT

Floricane-fruiting cultivars

The leader or crown bud will expand and produce canes. Canes will grow until short days and/or cold weather limit further growth. At the same time axillary buds start to change from vegetative to reproductive (flower) buds. This process of changing is called differentiation and is also described as the transition from a primocane to a floricane. For floricane fruiting cultivars, the first several buds (usually seven or eight) from the tip change first and then progress toward the base of the plant. At the base, several buds below ground level become leader or crown buds, which will produce the next year's primocanes. A vigorous cane and a mild autumn will produce more flowers per bud (node) than less vigorous canes. As canes stop growing, acclimation begins and as the leaves fall, the plant enters dormancy. Dana and Goulart (1989) in NRAES-35, suggest that low temperature is the critical factor in the change from a primocane to a floricane. This may take several weeks. The stage called complete dormancy is then entered and is not readily reversed until a specific number of chilling hours have been reached.

To overcome true dormancy, most raspberry cultivars require 800–1600 Richardson Chill Units (1 h at 0–6°C (32–42°F) equals 1 h of chilling), depending on cultivar and environmental conditions. According to Jennings (1988), red raspberries are influenced by growing season conditions, genetic difference of cultivar, and age of plant at the time that dormancy begins. Two-year-old plants may grow longer in the autumn than 7-year-old plants and their buds may start to grow earlier in late winter (Jennings, 1988). Not all buds have equal chilling requirements. Generally, flower buds have less of a chilling requirement than lateral buds and lateral buds have lower chilling requirements than terminal buds, and buds on thinner canes require less chill than buds on thicker canes of the same cultivar. In Canada, Dale indicates that raspberries are able to break dormancy in early January for greenhouse production (Dale *et al.*, 2003). In Peoria, Illinois (40°, 40 min latitude north) Skirvin *et al.* (1998), using data from 1951 to 1995, reported that 800 h of chilling are 90% complete by 2 February and 1600 h are 80% met by 30 April in field production. Thus, when the chilling requirement is met, the plant is ready to grow regardless of the time of year. Further, in areas having warm March weather followed by cold weather and frosts, flowers are damaged. According to Skirvin *et al.* (1998), this phenomenon would occur on raspberry cultivars having 1000 chilling h. In the US Midwest, the growing season may not begin until early May, after the last frost-free day. Northern hemisphere growers and researchers can gather and compare weather data and test for the breaking of dormancy by taking plants into a greenhouse in late December through February to determine the chilling requirement. Insufficient winter chilling may be more significant in Southern Europe or Australia where summer temperatures are very high. At these locations, canes have few laterals because their bud dormancy is not met. Thus, certain low-chill cultivars may be more productive.

Cold hardiness is another factor in winter plant survival and the development of leaves and flowers for the floricane. Red raspberry cultivars in the USA can withstand −28°C to −35°C (−20°F to −30°F) and black raspberry cultivars can withstand −20°C to −26°C (−4°F to −15°F) during the winter if the plant has not been overly fertilized with nitrogen and the management of soil moisture, soil fertility or mulching has been moderate. Further, both genetic factors and environmental factors, such as fluctuating temperatures, cause cultivars to differ in winter hardiness and winter injury (winter tip die-back to cane death). Cane survival in the winter is also higher if the disease anthracnose is controlled. Cold injury to plants does occur after dormancy requirements have been fulfilled with the onset of warm temperatures above 5°C–20°C (41°F–68°F) during late winter, which are then followed by temperatures below 0°C (32°F).

Once the raspberry plant receives warm temperatures (above 7°C or 45°F) after the chilling requirement has been met, the fruiting laterals produce leaves and flowers. If the cane is healthy, all buds will emerge at the same time except on the lower 30 cm of the plant (Fig. 2.3). Cultivated raspberries are self-fruitful (pollen is transferred within the same flower for fruit set) but in the wild, plants are often self-infertile and may show some cross infertility. New cultivars should be checked for self-fertility before commercial release.

Bees are responsible for the most transfer of pollen and pollination. Fruit size is due to the number of ovules that are fertilized and develop into fully expanded drupelets (Fig. 2.4). Pollination and fertilization are influenced negatively by lack of bees, cold, wet weather, nutrient status and disease. Boron deficiency can reduce fruit size and influence cane die-back (Barney and Miles, 2007). Red raspberry fruits ripen 21–43 days after first bloom and black raspberries ripen 27–40 days after first bloom in central Pennsylvania.

Once a berry is removed from the receptacle it is still alive and respires (releasing energy and heat) (Plate 2.2). The respiration rate of raspberries is the highest among temperate fruit, and so it has a shorter shelf life and quality. Further, respiration reduces the health components (vitamins A and C). To ensure that most antioxidants are retained, berries should be refrigerated within minutes of harvest and during transport to the consumer. Rapid cooling in conventional refrigeration with 90–95% humidity is important for any berries for sale. Controlled atmosphere (high CO_2 and low O_2) storage can be used but normally small growers move their berries very quickly and do not use controlled storage (Goulart, 1995). Modified atmosphere packaging allows the natural CO_2 to build up and reduces O_2 but this still needs to be improved for commercial use.

Primocane-fruiting cultivars

Primocane-fruiting raspberries are planted and managed in the same way as floricane-fruiting cultivars during the first year. If vigorous nursery plants are planted in early spring and managed with fertilizer and irrigation, some fruit can be harvested in autumn. In very warm locations or subtropical conditions, a second crop can be produced after the first fruiting. The delay until this crop can be harvested is affected by how low the canes are pruned after harvest. If pruning is done just below the zone that produced the first primocane crop, then another harvest may be achieved in as little as 12 weeks, but if pruned to the ground, the delay will often be 20 weeks or more. In fully subtropical conditions, many primocane-fruiting cultivars will continue to cycle, producing two to three crops a year for around 3 years, after which a dormancy period is essential for continued growth. However, with the cultivar 'Summit', this does not appear to be necessary and plants are reported to continually cycle.

In warm temperate regions, the canes of some cultivars may be pruned just below the cane section that primocane fruited in the first season's growth, and the following spring an early floricane crop is produced. In cooler locations or where hand pruning is not desired, the first season's canes are cut off in winter at 2–4 cm above the ground. During the second year, vegetative buds below ground create new primocanes that will flower and fruit. Productivity of these plants is correlated to cane diameter and cane height. Tall canes with a large diameter at soil level will give the highest yield. As a planting continues to produce from years 3 to 6, cane diameter and cane height may decrease for many reasons, such as plant competition, root rot or other diseases. In the eastern USA, irrigation is needed in most seasons from June through harvest to achieve optimal cane growth and fruit development.

Primocane or autumn-bearing canes are considered day-neutral plants. Flower bud initiation on primocanes takes place on long sunny days in the summer. Flower bud initiation continues from the top to the bottom of the cane. The top flower cluster is the smallest and the lowest one is the largest. Flowering is correlated to cane height, usually 40–50 nodes with the first flowers opening in late June ('Autumn Bliss' in Ohio) to mid-July, followed by the first fruits ripening in late July to mid-August, depending on cultivar. Days from flowering to first harvest, as with the floricane-fruiting types, are subject to significant environmental effects, but as temperatures are often cooler they vary from 4 to 10 weeks or more. Productivity is highly correlated with cane growth and with temperatures. An early spring season and a warm summer will produce early fruiting; the opposite will cause later fruiting. Most primocane-fruiting cultivars will continue to ripen fruit until frost in cooler climates; mid-October in the field and into late November if under a high tunnel in Ohio. In other locations and with improved cultivars, the complete harvest may be achieved before the first frost or cold weather occurs. For example in Bryansk, Russia, 'Autumn Bliss' achieves around 42% of the crop by first frost on 15 September but other newer cultivars achieve their complete harvest before this date (Kazakov and Evdokimenko, 2007). In the Pacific Northwest, 'Autumn Bliss' ripens 3 weeks ahead of 'Heritage' and has a greater harvested yield by the first frost in late September.

CLIMATE AND ENVIRONMENT

Climate, with regard to temperature, wind and weather, and environment (the soil, air and water quality), has an impact on the physiology of the raspberry plant. Many of these topics have been discussed in the previous paragraphs and will be again in Chapter 3. The raspberry needs full sunlight for cane growth, good yields and quality fruit and will be less productive in partial shade. Further, an individual cane will be more productive when exposed to sunlight through correct spacing at planting. As the plants multiply and grow taller, with large cane diameter at the soil level, removing small canes will expose other canes to more sunlight. The ambient air temperature from the beginning of the year to the end of the year has a profound effect on growth, development and the reproductive cycles of the plant. Temperature affects bud initiation, fruit quality, acclimation and the breaking of dormancy when the chilling requirement has been met. High soil temperatures have a negative effect on plant growth. Using organic mulches and irrigation will help reduce this problem. High winds will cause cane breakage; cold winds can cause desiccation and expose canes to cold injury. The use of a trellis for all types of raspberries can reduce wind injury, improve sunlight exposure, and increase plant canopy per unit of land while supporting canes upright for optimal harvest by people or machine (see Chapter 10). Roots require free oxygen and so fields that are flooded for long periods can cause plant injury and death. Plant growth and optimal yields are enhanced by 5 cm of water per week with periodic rainfall, but in most seasons, additional controlled watering at the root zone with drip irrigation is preferred.

MEDICINAL USES AND HEALTH BENEFITS

Medicinal uses of raspberries have been studied for centuries and their diverse medicinal properties – in roots, flowers, fruit and leaves – have been known since antiquity. In Ohio, ripe fruit were found to contain more antioxidants than immature fruit. Red raspberry seeds contained more phenolic compounds (ellagic acid content) than the pulp, but leaves had a higher content than seeds or pulp (Funt, 2003). Raspberry seeds (oil content) contain high amounts of vitamin E. Researchers continue to discover many important components for good health and disease prevention. Recently, medical research has shown berries being used to treat over 60 medical conditions, including certain cancers (black raspberry research at The Ohio State University for esophageal, oral and colon cancer), diabetes and arthritis and improving the immune system (Chen et al., 2006; Hecht et al., 2006; Stoner et al., 2007; Stoner, 2009). Current research involving phytonutrients (primarily polyphenolic and triterpenoid) of raspberries validates many of these uses. Most raspberry phytonutrients are antioxidants (vitamins A, C, E, calcium, potassium, zinc). Phytonutrient combinations may act in concert with additive and synergistic bioactivity.

The compositions of black raspberries differ from red raspberries in certain compounds (see Chapter 16). Black raspberries have been shown to have higher oxygen radical absorbency capacity (ORAC) (Plate 2.3), anthocyanin and dietary fiber than red raspberries. However, red raspberries have been shown to have higher amounts of vitamins A and C and slightly higher amounts of vitamin E than black raspberries (Funt, 2003). Year-to-year variations in fruit mineral content in primocane-fruiting 'Heritage' and ellagic acid content in 'Jewel' black raspberry appeared to be due to cold temperatures before and during harvest (Funt, 2003). Wang et al. (2003) indicated that in strawberry plants, high day/night temperatures (22°C/70°F) enhanced both phenolic and antioxidant capacity. Cooler temperatures reduced phenolic and antioxidant capacity. It appears that regions that have optimal temperatures at harvest may have a more consistent year-to-year level of compounds that benefit human health.

Fruit with high levels of anthocyanin and health-promoting properties may be more appealing to consumers. Anthocyanin pigments are affected directly and indirectly by the changes in acids and sugars. Foods high in antioxidant levels also contain high levels of phenolic compounds (Perkins-Veazie and Collins, 2001). In 2011, most people in the USA were eating only one-third to one-half of the recommended daily amounts of fruits and vegetables based on guidance from the USDA. Further, raspberries are being recognized as beneficial to health and longevity. However, deterrents to high consumption of raspberries lie in their high cost or cost per serving and seasonal fluctuating prices, which may exclude people in the lower or middle socioeconomic ranks from routinely purchasing them (Scheerens, 2001). Recent introduction of a 'berry smoothie' (ice, berry pulp and juice) by a large fast food restaurant chain may stimulate greater raspberry consumption but it comes at a higher price than carbonated drinks.

REFERENCES

Barney, D.L and Miles, C. (eds) (2007) *Commercial Red Raspberry Production in the Pacific Northwest*. PNW 598. A Pacific Northwest Extension publication.

Chen, T., Hwang, H., Rose, M.E., Nines, R.G. and Stoner, G.D. (2006) Chemopreventive properties of black raspberries in N-nitrosomethylbenzylamine-induced rat esophageal tumorigenesis: down-regulation of cyclooxygenase-2, inducible nitric oxide synthase, and c-Jun. *Cancer Research* 66, 2853–2859.

Dale, A., Sample, A. and King, E. (2003) Breaking dormancy in red raspberries for greenhouse production. *HortScience* 38, 515–519.

Dana, M. and Goulart, B. (1989) Bramble biology. In: Northeast Regional Agricultural Engineering Service (NRAES-35). Cooperative Extension. Cornell University, Ithaca, New York, pp. 9–18.

Funt, R.C. (2003) Antioxidants in Ohio berries. *Acta Horticulturae* 626, 51–55.

Goulart, B.L. (1995) Postharvest Handling of Brambles. Michigan Regional Fruit School, MSU Raspberry School (notebook) 1995, 38–43.

Hecht, S.S., Huang, C, Stoner, G.D., Li, J., Kenney, P.M., Sturla, S.J. and Carmella, S.G. (2006) Identification of cyanidin glycosides as constituents of freeze-dried black raspberries which inhibit anti-benzo[a]pyrene-7,8-diol-9,10-epoxide induced NFB and AP-1 activity. *Carcinogenesis* 27, 1517–1525.

Jennings, D.L. (1988) *Raspberries and Blackberries: Their Breeding, Diseases and Growth*. Academic Press, London.

Kazakov, I.V. and Evdokimenko, S.N. (2007) *Primocane Fruiting Raspberries*. All-Russian Selective-Technological Institute of Horticulture and Nursery Production of the Russian Agricultural Academy, Moscow.

Perkins-Veazie, P. and Collins, J.K. (2001) Contributions of nonvolatile phytochemicals to nutrition and flavor. *HortTechnology* 11, 539–546.

Scheerens, J.C. (2001) Phytochemicals and the consumer: Factors affecting fruit and vegetable consumption and the potential for increasing small fruit in the diet. *HortTechnology* 11, 547–556.

Skirvin, R.M., Otterbacher, A.G., Kunkel, K.E., Czubak, P. and Yiesla, S.A. (1998) Applications of a Chilling Hour Climatology to Predict Fruit Crop Growth in Illinois or How to Tell When Your Fruit Crops Are Ready to Start Growing in the Spring. Proceedings, Illinois Small Fruit and Strawberry Schools, University of Illinois Cooperative Extension Service Horticulture Series 102, pp. 54–64.

Stoner, G.D. (2009) Foodstuffs for preventing cancer: The preclinical and clinical development of berries. *Cancer Prevention Research* 2, 187–194.

Stoner, G.D., Wang, L.S., Zikri, N. and Chjen, T. (2007) Cancer prevention with freeze-dried berries and berry components. *Seminars in Cancer Biology* 17, 403–410.

Wang, S.Y., Zheng, W. and Maas, J.L. (2003) High plant growth temperatures increase antioxidant capacities in strawberry fruit. *Acta Horticulturae* 626, 57–63.

3

CLIMATIC REQUIREMENTS

HARVEY K. HALL[1]* AND TIM SOBEY[2]

[1]*Shekinah Berries Ltd, Motueka, New Zealand;* [2]*TSA Consultants, Herefordshire, UK*

INTRODUCTION

A study of the multifaceted relationship between raspberries and the growing environment has made it possible to identify some of the ranges of conditions that promote or inhibit the plant's ability to progress through the phenological stages of growth from juvenility to senescence. Each facet of climate has a range of effects on raspberries. Temperature has a primary effect on photosynthesis and then on growth, vigor, internode length, differentiation of plant organelles, speed of maturation, ripening and post-harvest quality. Wind also impacts growth, partly through physical buffeting but also through cooling effects via enhanced evaporation. Wind also brings pests and causes desiccation and water loss, especially at high or low temperatures.

Water availability is a critical factor for plant performance and causes significant stress if restricted or over-abundant in supply. The right amount of water is also essential for effective gas exchange within all parts of the plant, especially in the roots and leaves. With excess soil water, roots become starved of oxygen, and with insufficient water leaf stomata close, preventing gas exchange and limiting respiration.

A temperate climate is essential for almost all raspberry cultivars. For many raspberry species, chilling is required for seed germination and for floricane bud break.

CLIMATE ZONES

Raspberries are a crop adaptable to cool temperate climatic conditions. The natural range of the two primary species, *Rubus idaeus* and *Rubus strigosus*, or the subspecies *R. idaeus* var. *vulgatus* and *R. idaeus* var. *strigosus* (depending on

* hkhall@clear.net.nz

how they are divided taxonomically), is entirely in the northern hemisphere. At sea level wild raspberry species extend as far north as the Arctic Circle and at higher altitudes as far south as Mexico in North America, as far south as the mountains in Spain, Italy and Greece in Europe, and as far south as the mountains of Turkey, northern Iraq, Iran, Afghanistan, and through the mountains south of China across Asia Minor and Asia, right down into Australia. They can also be found on the equator in elevated regions of central Africa and in New Guinea.

Climatic factors, especially the cold temperatures in the north, limit the growth of the wild species. Winter kill of canes is a significant factor that limits the range in the northernmost extent of the species, especially in years with severe winters. After growth has begun in the spring, frost also causes damage to wild raspberries and prevents flowering, fruiting and dispersal of seeds in some years. Sudden cold, below $-5°C$ ($23°F$), before acclimation in the autumn can destroy the vascular systems and prevent flowering the following spring.

Approaching the equator, the limits of the range of spread of wild raspberry species are high temperatures, drought, high humidity accompanied by increased susceptibility to fungal diseases, and lack of winter chill. High temperatures limit fruit set at flowering time and reduce the quality of fruit through the action of sunburn and scald. Thus, most of the equatorial species are at higher altitudes where temperatures are lower and winter chill is greater.

Cultivated raspberries have been selected to extend the range of adaptability and are now grown over a much greater area than the wild species, including closer to the equator, at lower altitude and also around the southern hemisphere and even into the tropics at higher altitudes. Cold hardiness has been selected in the northeastern USA and Canada and throughout the Russian Federation, enabling the spread of commercial culture in regions where raspberries could not previously survive.

Floricane-fruiting types with winter hardiness are now grown throughout the northern hemisphere, although at the limits of this range fruit quality and other commercially important traits are not as well developed as cultivars grown in less demanding climates. Floricane-fruiting types have also been developed that require less winter chill and with specific disease resistances, enabling growth and commercial plantings in regions where temperatures and humidity are high, although again the adapted types have poorer production and fruit quality.

The development of primocane-fruiting raspberry cultivars has given a great impetus to the growth of the adaptive range of higher quality raspberry cultivars. Selection of primocane-fruiting cultivars has eliminated the requirement for cane cold-hardiness sufficient to survive winters in the cold continental climates of the Russian Federation, Canada (north of the Fraser Valley) and east of the Rockies and also the Midwestern and north-eastern USA. With some new cultivars developed in Russia, it has been possible to grow raspberries and produce a crop north of the Arctic Circle, even outdoors on the south side of dwellings. In regions close to the equator, primocane-fruiting cultivars thrive with less chill than floricane-fruiting types and with some cultivars, such as 'Summit', it is possible to continue to grow a commercial planting for years, getting around three crops per year, without the plants going dormant or suffering from lack of winter chill. When the first crop has been completed canes are cut off and a new crop ensues. Subsequent crops are produced in a similar manner.

Winter

Winter is both a challenge and a necessity for most raspberry cultivars, especially floricane-fruiting types. A period of cool or cold temperatures is required for the breaking of dormancy, both for germination of seeds (for breeding purposes) and for sprouting of fruiting buds of floricane-fruiting cultivars. Adapted cultivars need to have their winter cold requirement satisfied by the time the warmer temperatures of spring arrive. Closer to the equator and to the polar limits of raspberry cultivation, cultivars require less chill, whereas in the mid-latitudes, equivalent to Oregon and southern Washington on the west coast of the USA and New Jersey on the east coast, adapted floricane-fruiting types require more chill. In Europe, the line of higher chill requirement would appear to be through England, northern France, northern Italy and Serbia, but on the continent this is not as clear as in North America.

Less chill is available in regions closer to the equator due to the higher day and night temperatures. In regions closer to the poles in the highest chill areas, chill accumulation is limited by low temperatures as the plants lose their ability to register chill below somewhere around 0°C (32°F). The lower temperature of cessation of chill has not been clearly elucidated but it could vary slightly between genotypes, perhaps ceasing from the point at which cell metabolism is suspended by low temperatures.

Satisfying the chill requirement of floricane buds is essential for normal floricane bud break in the spring. When insufficient chill has been received, it is expressed in poor floricane bud break, and in extreme cases, total absence of growth unless plants are given extra chill artificially. The first sign of lack of chill is a longer 'cane leg' devoid of lateral growth, reduced bud break above the middle of the cane and/or shortened lateral growth in the cane base and above the cane middle. More severe lack of chill often results in the production of a tuft of laterals in the middle of the cane and another tuft at the top of the cane, or with more severe chill, the growth is only in the cane top. These symptoms are not only due to lack of chill, but they can also be caused by lack of roots due to physical root damage or the effects of a root pathogen or pest attacking this part of the plant. Particularly severe root damage can be caused by Phytophthora root rot, weevils or pests like 'grass grub' in New Zealand.

In addition, chill is usually required for root bud initiation and dormancy release of new primocane shoots. In warm climates, this may limit new cane numbers and also reduce the effectiveness of propagating by root cuttings. This was clearly visible in a US nursery in Washington State with roots supplied from California. When roots were produced in Washington State, bud break and shoot production was greater than roots of the same cultivars from California and the multiplication of canes in the nursery was significantly increased (H.K. Hall, n.d., personal communication).

In regions with colder winters, cane damage may be caused by low temperatures, especially below −20°C (−4°F). Early breeding work in the USA and Canada, especially in the north-east, focused a lot of attention on the development of cultivars with cold hardiness. A number of cultivars were developed that could be grown in this region, including 'Latham', 'Chief', 'Boyne', 'Ottawa' and 'Muskoka'. Some of these are still grown in this region and also in colder regions of northern and eastern Europe. Ongoing breeding for cold hardiness has only occurred in this latter region and considerable hardiness has been selected in some floricane-fruiting cultivars in the Russian Federation.

Cold hardiness is not only dependent on the innate ability of the cultivar to withstand cold, but also it requires the plant entering into dormancy effectively so that it becomes hardened to cold. In different locations, the onset of winter varies from a very quick and severe spike of cold to slow dropping of temperatures and gradual onset of cold conditions. Usually the onset of cold conditions is quicker and more severe in continental conditions, especially at higher latitudes. Selection of raspberry cultivars for cold hardiness is highly dependent on the environmental conditions in any specific locality and the hardiness requirements for very cold continental conditions are very different from maritime conditions or localities closer to the equator.

In continental conditions, the onset of winter is clear, as is the onset of spring, and there is little need for adapted plants to withstand periods of higher temperature during winter. Adapted cultivars enter dormancy early and can withstand severe cold, but frequently winter chill requirements are satisfied early in the winter and if a period of higher temperatures ensues then growth will often begin. If the cold conditions resume, then hardiness is often reduced and significant damage can occur.

In maritime conditions and at warmer locations, adapted cultivars need to enter a deeper dormancy, after which they can withstand periods of higher temperatures without bud break or growth. After spring is under way in these environments, an adapted cultivar will then begin to grow without having been damaged through periods of winter or early spring movement. This is particularly important in conditions in Oregon, southern Washington State and south of New Jersey in eastern USA and in England, southern France and other lower chill European locations. In Africa, Australia, New Zealand and South America, raspberry production is mostly in latitudes where there is little need for winter hardiness. In regions where there are fluctuating spring temperatures, there are significant benefits in growing floricane-fruiting type cultivars that require high accumulated heat units from bud burst to the first open flower, so that damage from late frosts following early bud break is reduced or eliminated.

Hardiness requirements of primocane-fruiting types are less exacting than floricane-fruiting types, and indeed, they do not require the cane winter hardiness of floricane cultivars if canes are removed after harvest or if they are not going to be cropped in the following spring or summer. This has made it possible for primocane cultivars to be grown in locations with very cold winters, and some cultivars with short stature can be cropped in a very short growing season where floricane raspberries would not either crop or survive.

Very cold winter conditions can also cause winter damage through dehydration of overwintering canes, especially when low temperatures are accompanied by wind. It is possible that artificial protection of canes could be done by spraying a protective coating of anti-transpirant onto canes, but this is not practiced commercially. However, in areas where there is sufficient snowfall, canes can be covered by snow to prevent dehydration. This is effective in eliminating damage from severe cold temperatures of below −20 to −30°C (−4 to −22°F).

In locations such as the Heilongjiang, Jilin and Liaoning provinces in north-eastern China, it is a common practice to lay down the floricanes of raspberries and bury them under approximately 15 cm (6 in) of soil. This is very labor intensive but is usually successful in protecting the canes from winter damage. While this is practiced on commercial cultivars in this region, the gathering of fruit from the wild raspberry *Rubus crataegifolius* is also practiced and these plants frequently escape winter damage. Much could be saved in terms of the inputs for production in this region if the cold hardiness of this species could be introgressed into cultivated types.

Apparent winter damage may also occur if precipitation over winter is high and roots are damaged by waterlogging and/or root disease, especially Phytophthora root rot. This may be alleviated through the use of drainage and by selection of well-drained soils for raspberry production.

Spring

Spring conditions suitable for the onset of growth of raspberries are mild temperatures and absence of frosts. Plants are sensitive to late frosts, below about −5°C (23°F), and when lateral growth has been sufficient for flower bud formation, frosts reaching temperatures below −1 to 2°C (28 to 30°F) are sufficient to cause damage. Severe frost damage in springtime can result from a frost of −8°C (18°F) on young primocanes. This type of severe winter damage has been experienced at East Malling Research (EMR) in the UK, causing damage to almost all their cultivars, selections and seedlings (H.K. Hall, n.d., personal observation).

When winter chill requirements of overwintering canes are adequately met, there is no severe cane damage either from cold over winter or other causes, and there are no root problems, bud break usually occurs regularly from the top to nearly the bottom (apical dominance and basipetal growth). Even when chill requirements are fully satisfied, bud break and growth occurs only in accordance with the amount of root volume. Small root volume results in bud burst from the top downwards, to the point beyond which the roots cannot support more. With growth under good soil fertility and the right spring conditions, flowering may be expected 6–10 weeks after bud break and fruiting 4–6 weeks later.

Good spring growth conditions are critical to the production of both primocane- and floricane-fruiting raspberry cultivars and adverse cold or windy conditions can result in losses from plant damage. In addition, conditions may favor pests or diseases that also may take their toll on production.

Summer

Like winter and spring, the conditions most suitable for raspberries are when there are no extremes, especially high temperatures. In raspberries, photosynthesis is most active at temperatures of around 17–21°C (63–70°F). Below 12°C (54°F), cool temperatures restrict growth; above 25°C (77°F), photosynthesis is significantly reduced and growth is limited.

The optimum temperature condition for raspberries has been reported at around 18°C (65°F), but it is clear that raspberries with a higher optimum may be selected from breeding populations grown at higher temperatures, especially through the introgression of genes from Asiatic species that grow in subtropical conditions (Stafne, 2000). The ability to withstand higher temperatures is extremely important for the future of raspberry production. Internationally, the emphasis is being placed more and more on production under glass or in tunnels to ensure quality and supply. Breeding and selection of new cultivars is central to the improvement of adaptation to high temperatures. Progress will be helped enormously through the application of technology alongside traditional plant breeding methodology. In addition to plant growth and plant health limitations, temperatures above 25°C (77°F) also cause reduction in pollen viability and poor fruit set.

While high temperatures limit growth and production of raspberries, extreme heat has a direct impact on fruit quality, causing sunburn in direct sunlight and scald even where fruit is hidden inside the bush (Renquist *et al.*, 1989; H.K. Hall, n.d., personal observation). Additionally, fruit firmness and coherence are reduced with heat; picking during high temperatures can result in fruits of poor structure and limited marketability. Sunburn, the production of white drupelets in intense sunlight and under high temperatures, is very important and often is a cause of significant economic losses when raspberries are grown on sunny days with high temperatures. The amount of sunburn also increases when there is a shower of rain and bright sunshine afterwards, especially when raindrops are left sitting on the fruit. Beginning overhead irrigation during sunny days with high temperatures can also cause sunburn in raspberry fruit exposed to sunshine.

High temperatures inside the bush, without the action of direct sunlight, appear to 'cook' the fruit on the bush, changing the fruit to a dull red color and very soft texture. The flavor of these fruits is very poor and they are unusable for either the fresh market or processing. In Australia, the onset of these conditions during periods of drought means that there may be no possibility of ameliorating temperature by using water. The damage is then often severe, and fruit ripening for at least 3 days after the high temperature conditions have ceased is often unmarketable. When water is used for misting, it is effective in reducing temperature and protecting both the plant and fruit. The effect of enhancing air movement to dissipate the heat from the fruit is unknown. When primocane-fruiting cultivars are grown through a period of high summer temperatures prior to fruiting, it is necessary to manage plant growth so that flowering and fruiting is optimal during the autumn. Both high temperatures and high light can have the effect of shortening the internodes of primocanes.

Under some conditions, when temperatures are above 40°C (104°F), leaves shrivel, dry up and blacken (leaf scorch) on the plant. When plants are under a water deficit, fruit and leaf damage is increased and losses are more significant. In contrast, when raspberry plants have a good root system and ample supply of water, the ability to withstand high temperatures is increased and losses are significantly reduced. Water relations also are implicated in retention of pollen viability and fruit set at higher temperatures.

Rainfall and high humidity during flowering increases susceptibility to fungal infections, which may develop later under suitable conditions, destroying fruit quality. In addition, high temperature and high humidity during harvest, even without prior infection and with very rapid cooling, can limit shelf life.

Autumn

Until an array of new primocane-fruiting cultivars was made available, autumn was a season of minimal inputs and maintenance in the fields for many raspberry growers. However, this period is critical for the production of healthy plants for fruiting the following spring and summer. Care is needed to control pests and diseases, and manage weeds, fertilizer inputs and watering, especially when rainfall is low.

In autumn, the onset of cooler weather and shortening days are key to flower initiation, with 'Tulameen' as an example of initiating flowers when temperatures have dropped to around 13°C (55°F) and day length shortened to around 14 h. If any attempt is made to dig, chill and grow canes of this cultivar for long cane production before flower initiation has taken place, the plants need to be planted under conditions that will provide the day length and temperature requirements to stimulate flower initiation, in order to avoid all growth being vegetative. If flower initiation needs to take place after transplanting of long canes, the flowering and fruiting is usually significantly delayed.

With the introduction of primocane-fruiting cultivars, autumn became another season for production of high-quality raspberry fruit. In regions with a Mediterranean-type climate, this period is often dry and humidity is low, making it necessary to irrigate several times a day to supply sufficient water to produce a good crop of sizeable, well-formed fruit. For raspberry productivity, plants need to be managed well during this period. Covering with tunnels is effective in extending the harvest season and maintaining fruit quality and shelf life, especially later in the autumn, and for protecting fruit when rainfall occurs.

WIND

Protection from wind is critical to healthy growth of raspberry canes. When plants are continually exposed to wind, new cane growth is short and plants do not prosper. In a field with a uniform soil type and minimal shelter over part of the field in Victoria, Australia, a 'Willamette' planting had growth up to 2 m (6 ft) tall where protected by windbreaks, but as short as 1 m (3 ft) in the exposed center of the field.

Wind also has dramatic effects on a fruiting field of raspberries, causing damage to fruit and to new canes, especially when the canes are spiny. Damage to unripe fruit causes rub-marks, which cause the fruit to become shrunken and distorted when it ripens. If a storm hits a crop with strong winds, damage can also include shredded leaves, broken laterals and broken tips in new cane growth. Fruit can also be damaged by strong winds, with drupelets punctured against spines and bruised by battering against each other. In addition, a strong wind can cause significant fruit losses through falling onto the ground in machine-harvested cultivars like 'Nootka', which has very low attachment strength when fully ripe. Fruit losses on outdoor crops due to wind are increased when the wind is accompanied by rainfall, especially heavy falls.

Combinations of wind and rain, even in the winter, can create conditions in which the cane movement at ground level can create holes in the soil much wider than the cane's diameter, into which rain can pour, and in excess, cause root damage. This can weaken the cane base and make it particularly vulnerable in unsupported young primocanes, especially when planting has been shallow and cane attachment to roots is weak.

RAINFALL

A climate with a regular rainfall, especially through the autumn, winter and early spring, is very beneficial for raspberries when they are planted on deep, well-drained soils. However, when excessive rainfall occurs, causing waterlogging and tractors and harvesting machines to get bogged down in wet soils, there may be major issues with the ability to harvest the crop, especially when soils are regularly cultivated. A significant rainfall event in the middle of harvest may also have dramatic effects on increasing fruit rots, sometimes resulting in the loss of half or more of the crop when the event is sustained over a period of a week or more during the peak harvest period of floricane-fruiting raspberries.

The effects of rainfall are many, from increasing sunburn after a brief shower on a sunny day, to increasing losses from wind, to causing fruit to become subject to fruit rots before harvest, and causing fruit that has been harvested wet to be prone to spoilage. Rainfall also helps generate an environment promoting leaf and cane diseases, and rainfall or dew is essential for the germination of grey mold spores on fruit, flowers and canes.

Lack of rainfall also affects raspberry production. The effects of drought impact raspberries very quickly as the plants are shallow rooted. These effects are compounded when the soil is light and has little water-holding capacity. When raspberries are planted on light soils, mulching with organic material is very beneficial and application of drip (trickle) irrigation several times per day is necessary for optimum growth. Use of these soils can be advantageous when raspberries are grown in tunnels and all irrigation and fertilizer is applied with drip irrigation.

Primocanes do not initiate flowering until the terminal bud is laid down. In ideal growing conditions, this may be much later than desired, especially if canes are to be dug and transported for long cane production. The period of active growth may be shortened by drought (water stress) at one or more stages during the plant growing period, stimulating flower initiation earlier. Water stress in raspberries can promote premature leaf drop, but can be used to advantage when long canes are dug in the late summer for bundling, refrigerated storage, transportation and out-of-season production.

HUMIDITY

Atmospheric humidity is important for growing raspberries, and management of humidity and available water is beneficial for control of plant health. High atmospheric humidity promotes the development of large fruit, and it is impossible to obtain large fruit in conditions of low humidity. In contrast, fruit grown in cool conditions with high humidity and cooler nights is large and of premium quality. However, if the day/night temperature difference is great, free water can form at night on all parts of the plant, which can encourage fungal infections to develop on fruits and canes. Fruit shelf life is limited with higher humidity, and this becomes more apparent and severe as humidity increases and diurnal temperature variation increases.

LIGHT

The effects of light on raspberry growth and physiological processes within the plant are profound. The duration of the light period, as day length – photoperiod, stimulates phenological change. For example, it acts as the primary environmental trigger for bud differentiation.

Light is essential to photosynthesis but the light intensity required for maximum photosynthesis in raspberries is low (Pritts, 2002), 1100 lumens (lux). More intense light on an individual leaf will not increase photosynthesis, but more intense light may increase the heat of the leaf to above the threshold of photosynthesis, thus allowing photosynthesis in spite of cool surrounding air temperatures. The reduction of light has potential to increase growth at higher temperatures, and the use of white plastic film or ultraviolet (UV) absorbing or reflecting films over tunnels in subtropical or tropical latitudes has considerable potential to improve returns from raspberries. It appears to be bringing the leaf temperature down to active photosynthetic temperatures, thus allowing significant increases in growth, flowering and fruiting.

Cane growth is upright or strongly ascending in the presence of surrounding light. However, when a row of raspberries is strongly shaded on one side by a hedge or some other light barrier, growth is frequently skewed towards the light and the plants all lean in that direction. Light also affects the growth of buds in canes that are bundled, and frequently buds that are tightly wrapped within a bundle do not break or grow. Buds that do grow, shoot out to find light and space away from competing laterals. This factor needs to be taken into account when raspberries are trained onto a V-trellis. If dormant canes are trained onto an open V, then bud break and lateral growth will fill the inside of the V. However, if the canes are trained onto two wires kept together, when bud break occurs and lateral growth begins, the growth is towards the outside. Subsequent separation of the two wires to produce a V results in most of the lateral growth being on the outside of the V, and fruit produced on these rows is predominantly on the outside of the V.

Light also hardens and thickens the leaf surface and stem epidermis and appears to be necessary for axillary buds to ripen. Under low light conditions, even at low temperatures, axillary buds may grow actively and produce chlorophyll-free, etiolated shoots.

On fruits, light can stimulate the epidermis to protect itself. Some raspberry cultivars, especially from Scottish origin, or descended from that material, develop a grey bloom where dense hairs are formed on the surface of the exposed or sunny side of the fruit. In extreme conditions, UV light associated with high temperatures can stimulate white drupelet (bleached drupelet) disorder. The skin of the drupelet is cauterized and the drupelet loses water content and color, with pigments being bleached and denatured.

FACTORS IN CLIMATE CHANGE

A simplistic view of climatic change would suggest that warmer climatic conditions will allow the extension of raspberry production into regions where previously colder winters have prevented cane survival. Closer to the equator, cultivars that were adapted previously are likely to become subject to poorer bud break and lower yields. However, it seems that one factor is becoming clear with changes in world weather. Whatever is happening to the world's climate, weather conditions are becoming more volatile, with more frequent conditions of 'freak' weather, with storms, high winds, heavy rainfall, droughts, heavy snowfall, hailstorms and extreme temperatures. To withstand these conditions, cultivars will need to become more resilient.

ADAPTING ENVIRONMENTAL CONDITIONS FOR RASPBERRY PRODUCTION

With the changes in marketing of raspberries from a predominantly process market to at least 50% of the sales now in the fresh market, there has been a strong move to adapt the growing environment for raspberry production. Raspberries have gone from being a field crop grown outdoors in cool temperate regions to the modern production environment, including production in pots (removal of soil issues and targeting of harvest dates), production in tunnels or under glass (removal of rainfall and weather issues such as wind, excessive cold and excessive radiation), use of shade (reduction of light and of high temperatures and wind), use of fertigation and use of drip irrigation (removing the reliance on soil nutrition supply). In addition, the growing regions have been extended through the use of long canes, primocane-fruiting types, and shortened crop cycles with only one or two crops from each planting in regions in the subtropics and tropics and even up to high altitude to gain specific desirable climatic windows for raspberry fruit production and supply year-round. These same techniques have increased the period over which raspberries can be harvested from one region to several regions, making a 4- to 5-month market supply possible. This can be accomplished on carefully selected sites.

REFERENCES

Pritts, M.P. (2002) From plant to plate: How can we redesign *Rubus* production systems to meet future expectations? *Acta Horticulturae* 585, 537–543.

Renquist, A.R., Hughes, H.G. and Rogoyski, M.K. (1989) Combined high temperature and ultraviolet radiation injury of red raspberry fruit. *HortScience* 24, 597–599.

Stafne, E.T. (2000) Leaf gas exchange characteristics of red raspberry germplasm, in a hot environment. *HortScience* 35, 278–280.

SITE SELECTION

Harvey K. Hall[1]* and Tim Sobey[2]

[1]*Shekinah Berries Ltd, Motueka, New Zealand;* [2]*TSA Consultants, Herefordshire, UK*

SOILS

Raspberries require deep, fertile, well-drained silty or sandy-loam soils that do not harbor pests, perennial weeds, parasitic plants, diseases from previous *Rosaceous* crops or members of the *Solanaceae* or residual chemicals, especially herbicides. Presence of organic material is particularly useful; it is a valuable buffer for retention of fertilizer and water, as well as providing plentiful aeration for good growing conditions and plant health, especially the roots. In addition, the organic matter contributes to good soil structure and makes the soil friable and easy to manage.

Soils with a high percentage of clay and poor drainage or a high water table during part of the year are not desirable. Areas that are prone to flooding should be avoided. In saturated soils, root respiration is impeded and ideal conditions are provided for the establishment of root pathogens, especially Phytophthora root rots.

Soils that have had their structure destroyed should be rehabilitated by growing cover crops and working them in to increase organic matter. Compaction pans that have been formed by cultivation or by traffic, especially during wet conditions, should be broken up by deep ripping and thorough cultivation in preparation for planting a crop of raspberries.

Nutrient supply in soils suitable for producing quality raspberries needs to be sufficient, without super abundance or deficiency so that growth is manageable and not restricted or limited by insufficient quantities of either macro- or micronutrients. The ideal conditions for raspberries has a 1 m depth of fertile, well-drained, well aerated soil at pH 5.5–6.5, free from water stress, either excess or shortage.

---

* hkhall@clear.net.nz

TOPOGRAPHY, EXPOSURE (ASPECT) AND LAYOUT

Flat or gently sloped land is good for raspberry production. It is very helpful if the slope and exposure of the field is similar across the whole field. In general, rows are planted north–south so that production on the rows is similar on each side of the row. Occasionally raspberries are planted on contour lines, but this is unusual. Soil conservation measures, if incorporated in the management protocols, are more commonly practiced by using other approaches such as growing a cover crop down the row to reduce washing of soil during rainfall.

In cold regions, where winter temperatures have potential to cause winter damage, air movement is beneficial. In regions where spring frosts are common, a slope allowing frost drainage is particularly good for reducing damage. Slopes with an exposure facing the equator are valuable in cool climate locations; however, in warmer regions an exposure facing away from the equator is useful for increasing winter chill and extending the dormant period. An earlier harvest is achieved in most locations when the exposure is facing the equator and the block receives greater heat unit accumulation. Exceptions to this occur when chilling is very limited and plants on slopes facing away from the equator are able to break bud and grow while others on slopes facing the equator remain dormant.

Field layout is enhanced significantly when blocks are square or rectangular, rather than being odd shaped to follow geographic features, such as streams, hills or other changes in soil type or exposure. This is particularly important when production is for fresh market and the plants are grown under protective structures. Farm layout, with all blocks in close proximity and near the packaging and cool storage facilities, is also valuable as it is easier to move fruit to the packing shed and into the refrigerated storage (cool store) from nearby fields. It is also valuable to have the blocks near the source of clean irrigation water and power supply for provision of irrigation and fertigation.

CLIMATE

A cool, Mediterranean-type, temperate climate with minimal wind, absence of storms, moderate summer and winter temperatures and lack of severe extremes of temperature is desirable for raspberry production. A continental climate is likely to be better than a maritime one because a smooth and definite transition from season to season is desirable. In maritime climates, variation in weather is much greater than in a continental climate and it is more difficult to grow raspberries effectively. However, the climate, moderated by cool offshore waters in coastal California, is very good for raspberry production in the Watsonville area and also further south near Oxnard. In regions with lower atmospheric humidity, fruit size may be reduced unless irrigated appropriately to optimize plant growth and productivity.

Very cold climates are quite restrictive to raspberry production. Floricanes can suffer significant winter damage due to excessive cold and severe temperature shocks. In north-eastern China, this is combated by burying the canes under 10–15 cm of soil and then digging out and training to trellising in the spring, when the cold season has passed. This methodology has been emulated in other areas with cold winters, where canes are laid down and covered by straw, allowing snow to cover them. Another means of combating the severe effects of winter cold is by growing primocane-fruiting cultivars that are cut to the ground after harvest. The crowns of the plants are able to survive the winter better than exposed floricanes.

In very hot conditions, heat can be reduced by sprinklers or mini-sprinklers underneath the plant canopy or in severe conditions by overhead sprinkling of the plants. Shading and the use of ultraviolet (UV) reflecting plastic films over tunnels also help protect growing or fruiting raspberry plants against hot conditions. Fogging also offers considerable assistance in reducing damage from heat and intense solar radiation.

Windbreaks are very important in regions with significant wind-run (see Appendix 1). The use of artificial shelter belts is also helpful in protecting plants from being battered by wind and from damage caused by abrasion and wind rub. Hot winds cause desiccation, quick dehydration of canes, leaves and fruit, and result in increased use of water and speedy drying out of plants without adequate irrigation. Windbreaks are essential in regions where storms are common, both to protect outdoor crops and especially to protect tunnels that are clad during the windy or stormy season. In planning a planting of raspberries in areas prone to wind, it is worth allowing time for the growth of shelter belts before planting to offer protection from the outset.

ELEVATION

The standard adiabatic lapse rate (reduction in temperature as elevation increases or increase in temperature as elevation decreases) is 0.6–1°C per 100 m (3–5°F per 1000 ft), depending on whether the air is moist or dry. This translates to later spring bud burst and increased time between the growth stages because of lower temperatures. However in some cases, elevation can increase chilling and bring forward the satisfying of chill requirements, making a quicker bud break in the spring possible and an earlier harvest. Variation in elevation affects seasonality, productivity and the zone of adaptation of raspberry cultivars.

The effect of elevation may be confused by the exposure, slope, soil type and proximity of hills and mountains nearby. In New Zealand and Chile, the cultivated land on which raspberries are grown is almost always on the valley floor. Moving up a valley in most cases results in closer proximity to mountains and a significant influence of air drainage. Just a few kilometers up a valley can result in a 2-week delay in harvest.

In Colombia, Peru and Venezuela, there is a great range in elevation from the coastal plains to high altitude. In these countries, it can be clearly seen that there are different climatic zones as one proceeds up into the hills and mountains, from tropical at sea level to temperate and cool temperate conditions, permafrost and glacier higher in the Andes. In a short distance, the same range of harvest season can be seen as is found in covering hundreds of kilometers at sea level. This can also be found in other tropical and subtropical locations around the world where higher elevations are found, including East Africa, India, Mexico, Indonesia and New Guinea and in some countries in the Middle East. In Mexico, there is considerable use of the different altitudinal climatic zones. There is some exploration of these different zones in other parts of the world, but there remains much to be explored and there is much research to be done.

ACCESS

Unless the site chosen has appropriate quality access to the chosen markets, essential infrastructure, the components of production, harvest and post-harvest, and the quality of soil and climate, it will not bring the appropriate rewards. Above all, an understanding of the markets that might be chosen, market periods of demand, the duration of those periods, their quality and delivery requirements, and the current and potentially available means of access should be thoroughly investigated before seeking sites to produce raspberries. Raspberry fruit is extremely perishable compared to other fruits, and so it is particularly important that fresh fruit can be transported to markets without loss of quality. Negotiating rough roads, transport over long distances and extended time taken to reach the marketplace reduce both fruit quality and duration of shelf life before sale and consumption.

SUPPLIERS TO THE RASPBERRY INDUSTRY

When a new farm or growing region for raspberries is under development, consideration must be given to access of suppliers of industry and horticultural products and services used in the growth and production of raspberries. These include, but are not limited to, suppliers of posts, wire, irrigation equipment, tunnels, fertigation equipment, pots, potting mix, fertilizer, spray chemicals, consultants for fertilizer, pest and disease management, growing and management expertise, specialists in picking, packing, quality control, refrigerated storage, cool chain, food safety standard licensing specialists, and inspection expertise for shipping and marketing in distant markets. For an operation to function efficiently, the employment of highly qualified staff and attention to detail are needed. Access is required to high-quality construction companies for development of packing facilities and refrigerated storage. Similarly, a high-quality electrical supply is required for running packing sheds and refrigerated storage, especially in tropical and subtropical conditions.

TRANSPORT

Modern and well-developed roads and transport systems are needed in close proximity to the raspberry production facility, and fields within a raspberry farm need to have high-quality access available under all weather conditions. Close proximity of the production fields to the refrigerated storage and packing sheds is desirable and for fresh market fruit it is essential, so that harvested fruit can regularly be moved into the cool chain. If access from the field to refrigerated storage and packing facilities is impeded, then even a short delay in getting harvested fruit into the refrigerated storage and getting field heat removal under way can result in loss of fruit quality and storage life of fresh market fruit is reduced.

If the roads within the farm or from the packing shed to the highway are rough, fruit quality is often reduced before it gets on the way to the market. Highway conditions also may not be conducive to maintaining fruit quality, especially if there are corrugations or the road bed has not been laid smoothly. Quality of fruit shipped from Los Reyes in Mexico to markets in the USA improved significantly when the road access from this area was improved. The key thing with raspberries in terms of their ability to survive physical damage from jolting or being dropped is that raspberry fruit rarely are firm enough to survive more than five drops before becoming unsaleable.

Problems with the transporting vehicles can also cause deterioration of fruit quality. If wheels are out of balance or there is vibration in the truck bed (deck), fruit will be affected, and in some cases the drupelet skin wears through, releasing juice. Wear also may occur in the walls of clamshells if they rub against each other, resulting in a white dust and a worn mark on the surface of the clamshell. Vehicles transporting raspberry fruit need to be refrigerated and if fruit is shipped in air cargo containers, it must be cooled and not be left sitting on the tarmac in the sunshine. The cool chain needs to be monitored and fruit kept at a constant temperature of 0–2°C (32–36°F) in the refrigerated transport, storage and to the market shelf.

LABOR

Both large-scale production of machine harvested raspberries for processing and a production unit of raspberries grown for hand harvest and fresh marketing require a large labor pool and reliable workforce. If the production block is established near a significant urban population, it can be advantageous both to the berry production unit and to the local labor pool. Providing transport can often be a useful tool for grower employers trying to attract a workforce. If the production unit is some distance from a metropolitan center with available workers, it may be necessary for the grower to supply housing for workers and also provide access to groceries and other supplies.

A key to harmonious working conditions is good labor management and a reliable picking tally each day. An essential part of this is effective harvest management, with fields kept neat and tidy, fruit being well presented, easily found and speedily harvested. Rows need to be assigned without undue competition, and it is important that harvest supervisors train pickers well and monitor picking thoroughness of each picker on a daily basis, so that quality issues will not ensue later in the harvest season. Rows should be an easy walking distance from weighing and accumulation sites in the field.

PACKING SHEDS

The site selection for packing sheds on a raspberry property is crucial to the efficient running of a production unit. The packing shed needs to be near the production fields so that fruit can be frequently and quickly taken from the field to maintain fruit quality, and so that field heat may be removed from the fruit as soon as possible after harvest. Access from fields to the packing shed needs to be on smooth, well-graded roads or tracks so that fruit will not be subjected to vibration or jolting en route. In regions where there is elevation on the property, locating packing sheds on an elevated site is valuable because this allows for constructing loading bays, for drainage of waste water, and for collection and storage of water from the catchment areas of roofs, hard surface access and parking areas. Access to a high-quality electrical supply to run the refrigerated storage and air conditioning units is also very important in choosing the site of a packing shed. Getting electricity supply to a site distant from supply lines may be a considerable expense.

With modern marketing and health standards, packing sheds are required to meet the health and the United Nations safety inspection criteria for marketing organizations, such as Hazard Analysis Critical Control Points (HACCP) and GLOBALGAP. They also need to be well constructed and designed for effective cooling, handling, packaging, storage and shipment of fresh or process berries. Packing sheds need to be designed to receive the fruit in an incoming area and move it quickly into a forced air cooling room and then into a holding area. Fruit is then passed through packing lines, where containers (clamshells, baskets or punnets) are given quality inspection, weighed and closed before packing into flats, stacking onto pallets and placing into the clean packaged fruit refrigerated storage to await dispatch.

Packing shed floors should be well constructed using insulated concrete, and the walls and roof built using metal-coated polystyrene refrigerated storage panels, with all surfaces easy to clean and keep hygienic. At the base of all panel walls, it is wise to install a protective buffer to reduce the risk of damage to walls through collision with pallets or forklifts. All forklifts inside the facility need to be electrically powered so that there are no fumes released into the packing shed. Internally, the different areas should be well isolated from one another through the use of both interior doors and draft screens. Cool temperatures (1–2°C, 34–36°F) should be maintained in each storage area and conditions inside the grading, checking, weighing and quality control area should be kept cooler than ambient conditions outside. Passage of fruit through the packing area and back into storage should be rapid so that fruit temperatures do not rise and result in loss of quality.

Packing sheds should be designed to provide staff facilities, such as toilets, meal and drink break areas, and sinks for washing so that hands are washed and feet kept clean for re-entering the packing area. No foot traffic should be allowed from soil outside into the packing area. Outside the facility, waste water needs to be connected to sewage mains or to on-site waste water treatment facilities so that there is no risk of disease or any form of contamination to the fresh fruit. Loading areas for both incoming and outgoing fruit should be on concrete or asphalt (bitumen) surfaces so that there is no dust or contamination coming into the packing shed. Providing a hard surfaced car park for workers who are walking into the packing shed is also worthwhile to reduce the likelihood of any contamination coming in on footwear.

PROCESSING PLANTS

When fruit is to be processed and hygienically packed for long-term storage or for retail packs, a range of further steps needs to be taken. Fruit passing through the processing plant needs to be assessed for microbiological contamination, particularly for the presence of communicable human disease. This is particularly important when the fruit is frozen, because microbiological contamination is often not sterilized but rather put into suspended animation by low temperatures.

To be able to monitor the microbiological status of fruit handled by processing plants that are freezing the fresh fruit, it is necessary to collect samples according to a regular testing program. These need to be placed onto various test media to show the presence or absence of pathogens.

If fruit is to be block frozen or frozen in drums, care must be taken to ensure that temperatures in the center of the containers reach freezing temperatures before bacteria and fungi proliferate.

For production of individually quick frozen (IQF) raspberries there are a number of additional factors that are important. First, cultivars that are to be frozen for IQF production need to be firm and have good fruit quality, so that they will not collapse during the freezing process. Secondly, the cultivars must be suited to the freezing process used. Only cultivars having fruit with small drupelets can withstand freezing with the use of liquid nitrogen, whereas cultivars frozen using the fluid bed technique are much less prone to freezing-induced bursting of drupelets, and they may have larger fruit and larger drupelets. Temperatures have to be carefully controlled so that fruit will not be damaged by mechanical impacts. 'Willamette' and 'Meeker' are both sensitive to temperatures being too cold, and if handled when fruit are at $-20°C$ ($-4°F$), are subject to shattering; fruit will crumble and quality is significantly reduced. At temperatures of $-16°C$ ($3°F$) losses of IQF fruit quality through crumbling are much lower.

If raspberries are processed by cooking, heating, juicing or evaporation, temperature monitoring is also very important. Heat treatment is able to kill any contamination by microorganisms but the temperature and duration has to be sufficient to destroy all bacterial and fungal growth. During the production of jams or preserves, it is necessary to constantly measure the temperature of the product to ensure that the heat treatment is sufficient to destroy all microorganisms.

MARKET

The most important factor to consider in setting up fresh or process raspberry production is the market – its proximity, its requirements and what is required to excel in the marketplace. The key for maximum returns, no matter how the raspberries are presented for marketing, is to deliver exemplary quality that is better than competitors and to maintain this quality throughout the production season for fresh fruit and year-round for process products. Delivery of quality fresh market fruit has been the key to market development by Driscoll's. They are quick to downgrade a product to a lesser brand if quality is lacking. Quality control is essential for every aspect of production and high-quality presentation is also obligatory. The best raspberry production operations have succeeded through the development of a robust consideration of marketing. Investment in raspberry production is worth considering when a market opening has been found and resources and expertise to deliver the high-end quality required for outstanding performance in that market have been acquired. A market analysis should include a detailed consideration of costs of production, including quality control, costs of packaging and refrigerated storage, costs of shipping and transport, costs of marketing, and a study of expected returns and the effects of supplying extra product into the market.

Key to the ability to deliver quality are the cultivars grown, the environment in which the plants are to be grown, control of environmental factors through the use of protective structures, such as windbreaks, tunnels or greenhouses, the use of cover crops, mulching or weed mats, and mechanical or hand weed control. When ripe fruit have been produced, optimum management of harvest labor, timely picking and rotation through blocks, rapid and gentle transit to and through refrigerated storage to packing sheds, to the market, and effective presentation and rotation of stock in the marketplace are essential in order to maximize returns.

In the modern market, retailers are keen to have high-quality fresh produce in their stores to attract buyers of other less perishable products. Fresh raspberries and other berries are key indicators to shoppers regarding freshness and quality as they are amongst the most perishable product lines. Thus, delivery of a great product is a win–win–win solution for producers, marketers and retailers, and when quality control is well managed, all will benefit from presenting high-quality fresh fruit to consumers.

Plate 1.1. (a) Fruit of *Rubus corchorifolius*, from China. **(b)** Cane of *R. corchorifolius*, from China. (Courtesy of Maxine Thompson and Judy Young.)
Plate 1.2. Fruit and leaves of *Rubus parvifolius*, from China. (Courtesy of Maxine Thompson and Judy Young.)
Plate 1.3. Fruit and leaves of *Rubus sumatranus*, from China. (Courtesy of Maxine Thompson and Judy Young.)

Plate 1.4. Fruit of *Rubus niveus*, from China. (Courtesy of Maxine Thompson and Judy Young.)
Plate 1.5. Fruit and leaves of *Rubus mesogaeus*, from Hokkaido, Japan. (Courtesy of Yuri Ito.)
Plate 1.6. Flower, young fruit and leaves of *Rubus hawaiiensis*, from Maui. (Courtesy of Joseph Postman.)
Plate 1.7. Fruit and leaves of *Rubus strigosus*, from South Dakota. (Courtesy of Michael Dossett.)

Plate 2.1. 'Cascade Delight', from the Washington State University breeding program at Puyallup, Washington, is a high-quality, large red raspberry for hand and mechanical harvesting. (Courtesy of David Karp, University of California.)

Plate 2.2. Black raspberry fruit separated from the receptacle. (Courtesy of The Ohio State University.)

Plate 2.3. Comparison of raspberries and blackberries for oxygen radical absorbency capacity (ORAC). (Source: Oregon Raspberry & Blackberry Commission.) Note: The ORAC fl (µmolTE/g) analysis, which utilizes fluorescein as the fluorescent probe, provides a measure of the scavenging capacity of antioxidants against the peroxyl radical, which is one of the most common reactive oxygen species (ROS) found in the body. Trolox, a water-soluble vitamin E analog, is used as the calibration standard and the ORAC result is expressed as µmol Trolox equivalent (TE) per gram.

5.1

5.2

5.3

Plate 5.1. 'Crimson Giant' fruit in a typical 150 g (6 oz) clamshell container, harvested in the winter in Morocco for the European market. The bright red color is desired for today's supermarkets and does not darken appreciably with storage. (Courtesy of C. Weber, Cornell University.)

Plate 5.2. Fully ripe fruit from 'Caroline'. Note the darker color of the closest fruit, which is the ripest. 'Caroline' continues to darken with storage and is not suitable for larger wholesale markets. (Courtesy of C. Weber, Cornell University.)

Plate 5.3. The variety 'Polka' is somewhat dark but the shiny skin makes it acceptable for wholesale markets, although not preferred. Note the fruit in the upper right of the cluster is overripe and has turned a flat dark red color. (Courtesy of C. Weber, Cornell University.)

5.4a **5.4b**

5.5

5.6

Plate 5.4. True yellow fruit color **(a)** differs from the amber or golden color **(b)**, as demonstrated in the fruit of these selections from the Cornell University breeding program. (Courtesy of C. Weber, Cornell University.)

Plate 5.5. 'Meeker' is the standard variety for freezing and processing for much of this market. Note the darker red color and slightly conic shape. (Courtesy of C. Weber, Cornell University.)

Plate 5.6. 'Jewel' black raspberry exhibits the typical fruit shape and color of *Rubus occidentalis*. The unique flavor profile and health benefits of black raspberries have driven a resurgence of interest in the North American native. (Courtesy of C. Weber, Cornell University.)

5.7

5.8

5.9

Plate 5.7. 'Tulameen' produces large conic berries with excellent flavor that are highly sought after in the marketplace. Unfortunately, low temperature sensitivity, susceptibility to *Phytophthora rubi*, low yield and a high chilling requirement have limited its production for commercial markets. (Courtesy of C. Weber, Cornell University.)

Plate 5.8. Many modern varieties are susceptible to root rot caused by *Phytophthora rubi*, which is exacerbated by poor drainage. Resistance is polygenic in nature but is present in varieties such as 'Prelude', 'Caroline' and 'Latham'. (Courtesy of C. Weber, Cornell University.)

Plate 5.9. Potato leaf hopper (*Empoasca fabae*) damage on developing primocanes of the variety 'Polka'. Note the chlorotic, crinkled leaves that give the plant the appearance of virus infection. (Courtesy of C. Weber, Cornell University.)

Plate 5.10. The large raspberry aphid (*Amphorophoro agathonica*) pictured and the similar small raspberry aphid (*Aphis rubicola*) are especially problematic in the spread of viruses that cause raspberry mosaic disease and others. (Courtesy of C. Weber, Cornell University.)

Plate 5.11. Symptoms of raspberry mosaic disease, which can lead to crumbly fruit and plant decline. This disease is spread by raspberry aphids and severely reduces yield and fruit quality. (Courtesy of C. Weber, Cornell University.)

Plate 5.12. Verticillium wilt in black raspberry displays a characteristic flagging of primocanes with a bluish tinge. (Courtesy of C. Weber, Cornell University.)

Plate 5.13. The spine-free canes of 'Joan J' **(a)** are popular with home gardeners and pruning crews compared to those of 'Crimson Giant' **(b)** and most other varieties. (Courtesy of C. Weber, Cornell University.)

Plate 6.1. Propagation of 'mother' plants in the nuclear stock facility at The James Hutton Institute. (Copyright The James Hutton Institute.)
Plate 6.2. *In vitro* propagation of raspberry. (Copyright The James Hutton Institute.)
Plate 6.3. Crumbly (left) and normal (right) fruit on raspberry. (Copyright The James Hutton Institute.)

Plate 6.4. Graft inoculation to *Rubus occidentalis*. (Copyright The James Hutton Institute.)
Plate 6.5. Sap indicators displaying symptoms after inoculation with tomato black ring virus (TBRV). (Copyright The James Hutton Institute.)
Plate 6.6. Cane death from raspberry root rot infection. (Copyright The James Hutton Institute.)

6.7

6.8

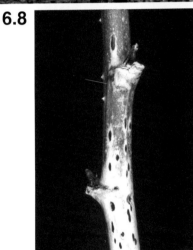

6.9

Plate 6.7. Irrigation spray stake prevents back siphon. (Copyright The James Hutton Institute.)
Plate 6.8. Black sclerotia on overwintered raspberry cane. (Copyright The James Hutton Institute.)
Plate 6.9. Damage to foliage caused by feeding by raspberry leaf and bud mite. (Copyright The James Hutton Institute.)

7.1

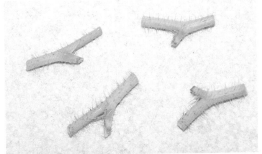

7.2

7.3

7.4

Plate 7.1. Nodal sections of raspberry canes are cut to approximately 2.5 cm (1 in) in length, containing one axillary bud per nodal section. These are sterilized and placed in growth medium to induce elongation of the buds. (Courtesy of C. Weber, Cornell University.)

Plate 7.2. The nodal segments are trimmed after sterilization to remove tissue damaged by the bleach and to provide a fresh cut surface for uptake of nutrients from the growth medium. (Courtesy of C. Weber, Cornell University.)

Plate 7.3. Nodes are placed upright in the growth medium supported by the agar (0.8%) for elongation. Note the red color from the Sprint®138 iron supplement. (Courtesy of C. Weber, Cornell University.)

Plate 7.4. Bud elongation and shoot development 10 days after placing nodal segments in the growth medium. The larger shoots will be ready for harvest within 5 days but can be left to elongate further. (Courtesy of C. Weber, Cornell University.)

8.1

8.2

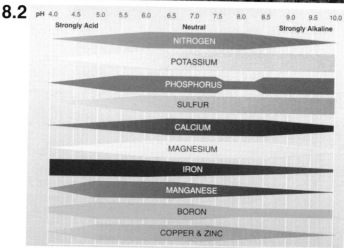

8.3

Plate 8.1. Raspberry field treated with lime after plants exhibited poor growth. This lime will be relatively ineffective at increasing pH in the root zone. (Courtesy of M. Pritts, Cornell University.)
Plate 8.2. Availability of soil nutrients varies with pH. (Courtesy of M. Pritts, Cornell University.)
Plate 8.3. Raspberry plant exhibiting iron chlorosis on young growth that was induced by alkaline soil. (Courtesy of M. Pritts, Cornell University.)

5

CULTIVAR DEVELOPMENT AND SELECTION

COURTNEY WEBER*

Department of Horticulture, Cornell University–New York State Agricultural Experiment Station, New York, USA

ORIGIN-REGIONAL ADAPTATIONS

The history of raspberry consumption and domestication goes back to ancient times (Roach, 1985; Daubeny, 1996), but the industry relied primarily on wild stands and propagation of superior chance seedlings until the late 1800s (Roach, 1985). Founding red raspberry varieties, such as 'Lloyd George', discovered in the early 1900s, and 'Cuthbert', discovered in the 1870s, still were chance seedlings found in the wild or in gardens (Hedrick, 1925). The growth in the current industry has accelerated the development of new varieties in the last 20 years yet heirloom varieties, such as 'Latham' (1912) and 'Taylor' (1935), are still commonly sold to homeowners in the USA. For a more extensive history of raspberry domestication see Roach (1985) and Jennings (1988).

Variety development in black raspberries, *Rubus occidentalis* L., is even less advanced than for red raspberries. A 2003 study on genetic diversity in black raspberries found that all commercially available varieties at the time were no more than two generations from a wild ancestor (Weber, 2003) and many were actually selected from the wild. Variety development in other raspberry species, such as the Arctic raspberry, *Rubus arcticus* L., and the Asian species *Rubus chingii* Hu, *Rubus crataegifolius* L. and *Rubus parvifolius* L., is even less advanced although significant effort is being expended in Korea on the native black raspberry, *Rubus coreanus* Miq. (Kim *et al.*, 2008). Many other raspberry species exist and are harvested in their wild state but have only made token contributions to commercial varieties through their use in breeding (Daubeny, 1996; Hall *et al.*, 2009).

---

* caw34@cornell.edu

BREEDING PROGRAMS

Fresh raspberry production has increased dramatically since the early 2000s. As recently as 2002, the value of red raspberries produced in the USA for processing exceeded the value of fresh market production, according to the US Department of Agriculture (USDA) National Agricultural Statistics Service (NASS, 2005) and fresh market production now exceeds process production in quantity as well as value. In 2010, the value of fresh market production was roughly four times that of processed raspberry production, which has remained relatively stable (NASS, 2011). The increase in fresh market demand sparked a renewed interest in raspberry breeding during this time, especially in private companies developing primocane-fruiting red raspberries. Numerous private breeding programs, including Driscoll Strawberry Associates (DSA) and Plant Sciences (both in Watsonville, California), Hortifruit Inc. (Santiago, Chile), Five Aces Breeding, LLC (Laurel, Maryland), Redbridge/Redeva (Peterborough, UK) and Shekinah Berries Ltd (Nelson, New Zealand and Queensland and Tasmania, Australia) are developing varieties for multinational production (Hall *et al.*, 2009, H.K. Hall, n.d., personal communication). Efforts at public institutions have been severely reduced in recent decades in the USA, with only four programs remaining. The USDA-Agricultural Research Service maintains a breeding program in collaboration with Oregon State University at Corvallis, Oregon, Cornell University has a program at the New York State Agricultural Experiment Station in Geneva, New York, Washington State University houses a program in Puyallup, Washingston and North Carolina State University maintains its program in Raleigh, North Carolina.

Outside the USA, federal and/or provincial governments, often in collaboration with private industry, sponsor most programs. In Canada, raspberry breeding is sponsored by Agriculture and Agri-Food Canada, primarily at the Pacific Agri-Food Research Center (PARC) in Agassiz, British Columbia (currently being discontinued), but also to a lesser degree at the Atlantic Food and Horticultural Research Centre (AFHRC) in Kentville, Nova Scotia. Raspberry breeding in the UK is centered at The James Hutton Institute (formerly the Scottish Crop Research Institute (SCRI)) in Dundee and at East Malling Research (EMR) in Kent. The Polish breeding program is located at the Fruit Experiment Station in Brzezna, while the Horticultural Research Station Centre in Polli houses a public program for Estonia. In Asia, the Korea Forest Research Institute in Suwon houses a public program focusing on the Korean black raspberry. In New Zealand, the raspberry breeding program became part of the New Zealand Institute for Plant and Food Research Limited in 2008, when HortResearch and Crop and Food Research merged. Other public breeding programs exist to some degree in Serbia, Hungary, Romania, Italy, Australia, China, Ukraine and Russia, but information on these programs is less available and their future status is unclear. Additionally, private breeding programs affiliated with other companies or individuals are known to exist in the USA, Australia, New Zealand, the UK, Mexico, Serbia, Chile and Spain, but little information is available on these programs (Finn *et al.*, 2008; Hall *et al.*, 2009; H.K. Hall, n.d., personal communication).

BREEDING GOALS

The development of new varieties hinges on a great many characteristics of the raspberry plant. However, the main goal must always be to improve on what is now available with adaptation to the particular growing location. This can have many definitions but the most common targets are improved fruit quality and higher yields. Disease and insect resistance, plant vigor, fruiting season, ability to machine harvest, adaptability to varying soil and climatic conditions and plant growth habit are all important traits to evaluate in potential new varieties, but must always be viewed in relation to the product, that is, high yields of flavorful berries. The wild-type red raspberry is small (1–1.5 g), soft and usually a dark, shiny red with a range of flavors from bland to outstanding. The ideal commercial variety produces high yields of large, firm, conical-shaped fruit with high sugars and a pleasant contrasting acidity. The ideal color for the fresh market is a bright, shiny red that does not darken in storage and for processing, a uniform, darker red. The challenge today is to transfer traits of interest, especially environmental adaptation, disease and insect resistance, from wild material and less desirable varieties into improved cultivars, leaving deleterious traits behind. Tables 5.1 and 5.2 describe the major strengths and weaknesses of many modern varieties with regard to their utility in breeding new varieties.

Fruit quality characteristics including fruit size, color and/or pigment content, sugar content, acid content, firmness, shape, coherence and overall flavor are the most commonly selected traits. Fruit size in the largest, new varieties, such as 'Crimson Giant' (Plate 5.1) and 'Maravilla', often exceed 7 g for individual fruits and can average more than 5 g over an entire season. Fruit size is a combination of drupelet size and number and has been increasing as new varieties are developed. Fruit size has implications on yield and harvest efficiency because if the number of fruit is equal between two varieties, larger fruit can fill containers faster and will result in greater overall yield and reduced harvest costs. Also, larger fruit tend to be more appealing to the consumer, making fruit size an important trait for marketability.

Equally important to marketability is fruit color and/or pigment content. For the fresh market, lighter colored fruit with a shiny appearance are desired. These fruit maintain a fresh look longer than darker red fruit, especially under the artificial lighting found in supermarkets. Varieties such as 'Glen Ample', 'Glen Lyon', 'Himbo Top', 'Maravilla', 'Encore' and 'Crimson Giant' (Plate 5.1) maintain a bright red color even after cold storage. Other varieties such

Table 5.1. Commercial floricane varieties for the fresh market and processing industry.

Variety name	Origin	Year of release	Plant patent/ PVR[a]	Primary use	Major strengths	Major weaknesses
Red varieties						
'Adele'	The Plant and Food Institute of New Zealand Ltd	2006	PP# 20,773 PVR	Fresh	Very shiny bright red, large fruit	Needs greater fruit firmness and fruit coherence
'Awaroa'	The Plant and Food Institute of New Zealand Ltd	2006	PP# 20,746 PVR	Fresh	Very early production, dual cropping	Moderate harvest ease
'Boyne'	Morden Experimental Farm, Manitoba, Canada	1960	None	Fresh	Early production, cold hardy	Small, soft fruit, fire blight susceptible
'Canby'	USDA-ARS, Oregon, USA	1953	None	Fresh	Excellent eating quality, functionally spine free	Susceptible to cold, *Phytophthora* root rot and powdery mildew, small fruit
'Cascade Bounty'	Washington State University, USA	2005	PP# 17,985	Processing	*Phytophthora* root rot resistant, machine harvestable	Marginal winter hardiness
'Cascade Dawn'	Washington State University, USA	2005	PP# 18,246	Fresh	Early production, large fruit	Soft fruit
'Cascade Delight'	Washington State University, USA	2003	PP# 14,522	Fresh	*Phytophthora* resistant, large firm fruit	Susceptible to RBDV and aphids
'Cascade Nectar'	Washington State University, USA	2003	None	Processing	Excellent flavor, suitable for wine	Dark, soft fruit; susceptible to RBDV and aphids
'Chemainus'	PARC, Agriculture Agri-Food Canada, British Columbia	2003	PVR	Fresh/processing	Large, firm fruit; machine harvestable, suitable for IQF	Susceptible to RBDV
'Coho'	USDA-ARS, Oregon, USA	2001	None	Fresh/processing	Late season; large fruit size; may be machine harvested for IQF	Very late for IQF market; very susceptible to *Phytophthora*

'Cowichan'	PARC, Agriculture Agri-Food Canada, British Columbia	2001	PVR	Processing	Machine harvestable, firm fruit, RBDV resistant, winter hardy	Moderate yield, excessive growth, susceptible to Phytophthora root rot
'Encore'	Cornell University, New York, USA	1998	PP# 11,746	Fresh	Late production, large size, high-quality fruit	Phytophthora root rot susceptible
'Esquilmalt'	PARC, Agriculture Agri-Food Canada, British Columbia	2003	PVR	Fresh	Large, late fruit	Susceptible to RBDV, Phytophthora and cold
'Glen Ample'	SCRI, UK	1994	PVR	Fresh	Large, firm fruit	Susceptible to Phytophthora and cold, pale color
'Glen Lyon'	SCRI, UK	1985	PP# 11,418 None	Fresh	Large firm fruit	Pale color, average flavor
'Jeanne D'Orleans'	Agriculture and Agri-Food Canada, St-Jean-sur-Richelieu, Quebec	2007	PP# 20,105	Fresh	Large fruit size, good shelf life and fruit quality for Quebec	Dark red fruit
'K81-6'	AAFC, Agriculture Agri-Food Canada, Nova Scotia	~1990	None	Fresh	Late production, very large fruit	Fire blight susceptible, soft fruit
'Killarney'	Morden Experimental Farm, Manitoba, Canada	1961	None	Fresh	Early production, cold hardy, bright red color	Small, soft fruit
'Korere'	The Plant and Food Institute of New Zealand Ltd	2006	PP# 20,772 PVR	Fresh	Early production, ease of harvest, spineless	Small fruit
'Korpiko'	The Plant and Food Institute of New Zealand Ltd	2006	PP# 20,771 PVR	Fresh	High yield, firm fruit, excellent quality	Root rot susceptible
'Latham'	University of Minnesota, USA	1912	None	Fresh	Phytophthora root rot resistant, vigorous	Small, soft fruit
'Lauren'	University of Maryland, USA	1997	PP# 10,610	Fresh	Large, high-quality fruit	Phytophthora root rot susceptible, cold susceptible

Continued

Table 5.1. Continued

Variety name	Origin	Year of release	Plant patent/ PVR[a]	Primary use	Major strengths	Major weaknesses
'Malahat'	PARC, Agriculture Agri-Food Canada, British Columbia	2000	PVR	Fresh	Early, high fruit quality, machine harvestable	RBDV susceptible, soft fruit, very susceptible to Phytophthora root rot
'Mandarin'	North Carolina State University, USA	1955	None	Fresh	Heat tolerant	Cold susceptible, poor fruit quality
'Meeker'	Washington State University, USA	1967	None	Processing	High color, machine harvestable	Phytophthora root rot, RBDV susceptible
'Motueka'	New Zealand Institute for Plant and Food Research Limited	2000	PP# 14,035 PVR	Processing	Excellent machine harvest ability	Root rot susceptible, leafy, medium vigor
'Moutere'	New Zealand Institute for Plant and Food Research Limited	2005	PP# 17,744 PVR	Fresh	Large, firm fruit, high yield; RBDV resistant	Poor flavor, flat red color, hard to pick
'Nanoose'	PARC, Agriculture Agri-Food Canada, British Columbia	2005	PVR	Fresh	High yield, large fruit, machine harvestable	Susceptible to Phytophthora root rot, RBDV and winter damage
'Nova'	AAFC, Agriculture Agri-Food Canada, Nova Scotia	1981	None	Fresh	Cold hardy, climate adaptability	Variable performance, medium small fruit
'Octavia'	EMR, UK	2002	PVR	Fresh	Late production, aphid resistant, large fruit	Poor shape, Phytophthora root rot susceptible
'Prelude'	Cornell University, New York, USA	1998	PP# 11,747	Fresh	Early production, disease resistant, double crop	Small fruit size, moderate yield
'Rudi'	PARC, Agriculture Agri-Food Canada, British Columbia	2010	PVR	Processing	High yield, machine harvestable, early	Susceptible to Phytophthora root rot and RBDV
'Sannich'	PARC, Agriculture Agri-Food Canada, British Columbia	2005	PVR	Fresh	High yield, machine harvestable	Fruit with low Brix and high acidity, medium fruit size

Cultivar	Origin	Year	Patent/PVR[a]	Use	Strengths	Weaknesses
'Tadmor'	The Plant and Food Institute of New Zealand Ltd	1999	PP# 14036 PVR	Fresh	Late season, high yield, high quality	Moderately hard to remove
'Taylor'	Cornell University, New York, USA	1935	None	Fresh	Vigorous plant, excellent flavor	Small fruit size
'Titan'	Cornell University, New York, USA	1985	PP# 5,404 expired	Fresh	Very large fruit, late production	Phytophthora root rot, average flavor
'Tulameen'	PARC, Agriculture Agri-Food Canada, British Columbia	1989	None	Fresh	Late production, high-quality fruit	Cold susceptible, Phytophthora root rot susceptible, low plant vigor
'Ukee'	PARC, Agriculture Agri-Food Canada, British Columbia	2010	PVR	Fresh, processing	Phytophthora root rot resistant	Moderate yield, light fruit color
'Wakefield'	New Zealand Institute for Plant and Food Research Limited	2008	PP# 21,185	Processing	Large sturdy canes; firm, dark fruit	RBDV susceptible; medium fruit size
'Willamette'	USDA-ARS, Oregon, USA	1943	None	Processing	RBDV resistant, machine harvestable	Moderate yield
Black and purple varieties						
'Bristol' (black)	Cornell University, New York, USA	1934	None	Fresh	Early production	Small fruit size, weeping form
'Brandywine' (purple)	Cornell University, New York, USA	1976	None	Processing	High vigor, good flavor	Flat purple fruit color, soft fruit
'Jewel' (black)	Cornell University, New York, USA	1973	None	Fresh	High-quality fruit	Moderate yield, weeping form
'Mac Black'	Anonymous private breeder, Michigan, USA	≈2000	None	Fresh	Late production, upright canes	Moderate yield and fruit size, poor color for black
'Munger' (black)	Ohio	1897	None	Fresh/processing	Machine harvestable	Virus susceptible, moderate yield
'Royalty' (purple)	Cornell University, New York, USA	1982	PP# 5,405 expired	Processing	Large fleshy fruit, aphid resistant	Flat purple fruit color, soft fruit

[a]PVR stands for plant variety rights, which are used in Europe, Canada and South Africa in a similar manner to plant patents in the USA.

Table 5.2. Primocane varieties for the commercial market.

Variety name	Origin	Release year	Plant patent/PVR[a]	Primary use	Major strengths	Major weaknesses
'Amira'	Berryplant, Verona, Italy	2009	PVR	Fresh	Very firm fruit, good shelf life	Low yield
'Anne'	University of Maryland, USA	1998	PP# 10,411	Fresh	Light yellow color, excellent flavor	Low cane production; low yield
'Autumn Bliss'	EMR, UK	1984	PP# 6,597 expired	Fresh/process	High vigor, large fruit	Dark, soft fruit
'Autumn Britten'	EMR, UK	1995	PVR	Fresh	Large fruit, early production	Sparse canes, dark red fruit
'Autumn Treasure'	EMR, UK	2008	PP# 20,769 PVR	Fresh	Spineless, upright, large conical fruit, easy to remove	Heat intolerant, drupelet wall thin
'Brice'	EMR, UK	2005	PVR	Fresh	Early production, spine free	Dark color
'Caroline'	University of Maryland, USA	1999	PP# 10,412	Fresh/process	High fruit quality, disease resistant	Overly vigorous
'Chinook'	USDA-ARS, Oregon, USA	2002	None	Fresh	Early production, large fruit	Questionable winter hardiness
'Crimson Giant'	Cornell University, New York, USA	2011	PP pending	Fresh	Very large fruit, late production	Moderately *Phytophthora* susceptible, late primocane production
'Driscoll Cardinal'	Driscoll Strawberry Associates, California, USA	2002	PP# 14,903 proprietary	Fresh	Firm fruit, consistent size	Not publicly available
'Driscoll Carmelina'	Driscoll Strawberry Associates, California, USA	2002	PP# 14,761 proprietary	Fresh	Late primocane fruit, firm, easy to remove	Not publicly available

Cultivar Development and Selection 63

'Driscoll Dulcita'	Driscoll Strawberry Associates, California, USA	2002	PP# 14,904 proprietary	Fresh	Firm fruit, good flavor and yield	Not publicly available
'Driscoll Estrella'	Driscoll Strawberry Associates, California, USA	2008	PP# 19,137 proprietary	Fresh	Yellow fruit, firm, high yield	Not publicly available
'Driscoll Francesca'	Driscoll Strawberry Associates, California, USA	2002	PP# 14,860 proprietary	Fresh	Firm fruit, good flavor and yield	Not publicly available
'Driscoll Madonna'	Driscoll Strawberry Associates, California, USA	2002	PP# 14,781 proprietary	Fresh	Large fruit, excellent firmness and flavor	Not publicly available
'Driscoll Pacifica'	Driscoll Strawberry Associates, California, USA	2008	PP# 18,658 proprietary	Fresh	Large red fruit, high yield, good flavor	Not publicly available
'Driscoll Sevillana'	Driscoll Strawberry Associates, California, USA	2006	PP# 18,659 proprietary	Fresh	Yellow fruit, firm, high yield	Not publicly available
'DrisRaspOne' ('Driscoll Ambrosia')	Driscoll Strawberry Associates, California, USA	2008	PP# 19,656 proprietary	Fresh	Large plant size, high productivity, large fruit, medium red	Not publicly available
'DrisRaspTwo'	Driscoll Strawberry Associates, California, USA	2011	PP# 22,246 proprietary	Fresh	High productivity, large fruit, dark red color	Not publicly available
'Elegance'	Plant Sciences Inc., California, USA	2009	PP# 21,685	Fresh	Firm glossy fruit, excellent flavor	Not publicly available
'Erika'	Centro Di Ricerca per La Frutticoltura, Rome, Italy	2008	PP# 20,841 PVR	Fresh	Large firm fruit, good color and flavor	Not publicly available

Continued

Table 5.2. Continued

Variety name	Origin	Release year	Plant patent/ PVR[a]	Primary use	Major strengths	Major weaknesses
'Grandeur'	Plant Sciences Inc., California, USA	2008	PP# 20,459	Fresh	Firm glossy fruit, excellent flavor	Not publicly available
'Heritage'	Cornell University, New York, USA	1969	None	Fresh/ process	Highly adaptable, upright canes	Small fruit, average flavor
'Himbo Top' (cv. 'Rafzaqu')	Promo-Fruit Ltd, Switzerland	2003	PP# 19,512 PVR	Fresh	Large, bright red fruit	Low cane production; leggy plants need extra trellising, soft fruit
'Isabel'	Driscoll Strawberry Associates, California, USA	1994	PP# 9,340 proprietary	Fresh	Large size, bright red color	Not publicly available
'Jaclyn'	University of Maryland, USA	2005	PP# 15,647	Fresh	Excellent flavor, large fruit	Dark color, fruit adheres tightly, potato leaf hopper susceptible
'Joan Irene'	Medway Fruits, UK	2005	PVR PP# 17,986	Fresh	Firm fruit, light red, spine free	Not publicly available
'Joan J'	Medway Fruits, UK	2002	PVR PP# 18,954	Fresh	Very productive, spine free	Dark red, excessive firmness
'Josephine'	University of Maryland, USA	2001	PP# 12,173	Fresh	Late production, excellent flavor	Dark red fruit
'Maravilla'	Driscoll Strawberry Associates, California, USA	2004	PP# 14,804 proprietary	Fresh	Large size, bright red color	Not publicly available, overly vigorous, very late harvest season
'Marcela'	Medway Fruits, UK	2005	PP# 17,819 proprietary	Fresh	Excellent shipping quality	Not publicly available
'Marcianna'	5 Aces Breeding LLC, Maryland, USA	2007	PP# 21,007 PVR proprietary	Fresh	Large, bright color, firm fruit, good flavor	Not publicly available

Cultivar Development and Selection 65

'Nantahala'	North Carolina State University, USA	2008	PP# 20,689	Fresh	Earlier, larger and firmer than Heritage	Low–medium productivity
'Pacific Deluxe'	Pacific Berry Breeding, California, USA	2008	PP# 21,074 proprietary	Fresh	Bright color, excellent shelf life, root rot resistant	Not publicly available
'Pacific Royale'	Pacific Berry Breeding, California, USA	2009	PP# 21,536 proprietary	Fresh	Large, firm, conical fruit, exceptional flavor, root rot resistant	Not publicly available
'Polana'	Research Institute of Pomology and Floriculture, Poland	1991	PVR	Fresh	Early production	Small fruit, extra fertilization needed
'Polka'	Research Institute of Pomology and Floriculture, Poland	2001	PVR	Fresh	Early production, large glossy fruit	Fruit darkens with storage, potato leaf hopper susceptible
'Radiance'	Plant Sciences Inc., California, USA	2008	PP# 20,342	Fresh	Firm glossy fruit, excellent flavor	Not publicly available
'Sugana'	Lubera AG, Switzerland	2008	PP# 21,357 PVR	Fresh	Very large fruit, excellent color	Not publicly available
'Summit'	USDA-ARS, Oregon, USA	1989	None	Fresh	Early season production, high yields, *Phytophthora* resistant	Small fruit size, inconsistent fruit quality

[a]PVR stands for plant variety rights, which are used in Europe, Australia, Canada, New Zealand and South Africa in a similar manner to plant patents in the USA.

as 'Canby', 'Killarney', 'Caroline' (Plate 5.2), 'Joan J', 'Autumn Britten' and 'Heritage' show significant darkening during storage. Dark red color can be offset somewhat by shiny skin as found in the variety 'Polka' (Plate 5.3), but this only delays the perception of over-ripeness for a short while. Most varieties developed prior to the 1990s are too dark by today's wholesale standards for fresh market production. The market for yellow (Plate 5.4a) and amber (Plate 5.4b) fruit is very small and few varieties of consequence are grown for most markets. 'Anne' produces some of the highest quality fruit in this category with fruit that is the completely recessive pale yellow with no blushing. Other varieties marketed as yellow include 'Kiwi Gold' and 'Golden Harvest', and are actually amber or blushed in color when fully ripe. Numerous yellow or amber varieties, including 'Golden Dome', are available in the Russian market but are unknown to Western consumers.

For the processing market, a darker fruit is desired both for frozen fruit and for further processing. Studies have shown that increased anthocyanin (the primary pigment in raspberry) levels provide greater color stability in processed products (Boyles and Wrolstad, 1993; Garcia-Viguera et al., 1999). The industry standard for frozen and processed raspberries is 'Meeker' (Plate 5.5), with other varieties, such as 'Willamette', 'Coho', 'Cowichan' and 'Cascade Bounty', also being grown for this market.

Black raspberries are prized for their high pigment content, which accounts for their extremely high antioxidant levels (Weber et al., 2008). The variety 'Munger' dominates the processed market while 'Jewel' (Plate 5.6) is the predominant variety for the fresh market. Most black raspberry cultivars have a shiny appearance on individual drupelets. This trait is not common to all black raspberries and in the wild a flat or dusty look to the berries is common. This is caused by microscopic hairs and waxy exudates that hold the drupelets together. Fruit firmness in black raspberries depends on a balance of these traits. The firmest types have high levels of hairs but are generally unattractive while a variety such as 'Haut' has very dark shiny berries with low levels of hairs. Its fruit is very soft and, while very attractive, does not hold well in a container.

The 'black × red' raspberry hybrids, marketed as purple raspberries, have an intermediate color that is not acceptable for the fresh market, except for local sales. All current varieties have the flat, dusty appearance that relegates them to local sales, primarily for home processing. However, breeding for shiny purple fruit is possible and may lead to an expansion of the color pallet for fresh raspberries in the future. Breeding selections in the Cornell University program have shown the shiny appearance to be possible in purple raspberries. Improvements in yield and flavor are needed to realize this potential.

Breeders and consumers alike consistently list fruit flavor as one of the most important characteristics that they look for in fruit. Unfortunately, the reality often falls far short of the ideal. Consumers consistently 'buy with their eyes' and breeders cater to this habit by producing large, pretty fruit that may or may not taste good. The persistence of popular fruit in the supermarket, such as 'Red Delicious' apples, winter strawberries from California, peaches and plums, is a testament to this phenomenon. Most raspberry varieties contain a fairly high level of flavor due to their high acid content combined with relatively high sugar content. Floricane varieties, such as 'Tulameen' (Plate 5.7) and 'Canby', and primocane varieties, such as 'Josephine', 'Jaclyn' and 'Anne', are considered to have outstanding flavor. Newer varieties, such as 'Crimson Giant' and 'Maravilla', can have very good flavor, but the ability to harvest these varieties prior to full ripeness tends to inhibit full flavor development and leads to overly acid or sour fruit. Some varieties, such as 'Haida', 'Titan' and 'Moutere', never produce better than acceptable flavor, regardless of their stage of ripeness.

For fresh market raspberries, shelf life has become one of the most important characteristics of new varieties. This 'trait' is actually a combination of fruit firmness, skin toughness, ability to pick early, fruit shape and drupelet coherence. Fruit firmness can be a measurement of the actual firmness of the flesh, such as in 'Joan J' and 'Moutere', which have extremely firm flesh that can seem rubbery at times, or the ability of the fruit cavity to resist collapse, as in 'Crimson Giant' and 'Heritage'. Fruit with very large cavities often collapses when stacked in containers. The shape of the fruit can influence this factor, with long conical fruit often resisting collapse better than broader fruit, again possibly a factor of the cavity size. Coherence is a measure of how well the drupelets hold together and is often associated with a uniform collar around the cavity opening. A non-uniform collar often leads to damage during picking and poor fruit quality. Finally, the ability to harvest early may be the greatest factor in shelf life. Some varieties, such as 'Meeker', 'Titan', 'Jaclyn' and 'Josephine', are very difficult to harvest until the fruit is completely ripe. This is the natural, wild type trait that ensures the seed matures prior to distribution by animals. However, it greatly reduces shelf life and increases the number of damaged fruit at harvest, because greater force is needed to pick fruit that looks ripe but does not release easily. New varieties, such as 'Crimson Giant', 'Maravilla' and 'Glen Ample', can be harvested when the fruit is pink or even slightly white, very firm and less prone to damage.

Disease and insect resistance remain primary targets for variety improvement in raspberries, although experience shows us that, in terms of commercial value, fruit quality and yield are more important. Many industry standards have serious disease and/or insect susceptibilities, yet continue to be grown because their fruit quality has not been matched. 'Meeker', the leading processing berry, is susceptible to the raspberry bushy dwarf virus (RBDV) that causes crumbly fruit. 'Tulameen', 'Encore', 'Canby', 'Titan', 'Lauren', 'Cowichan', 'Esquimalt', 'Autumn Byrd', 'Polana' and 'Himbo Top' are all susceptible to Phytophthora root rot, causing plantings to die out (Plate 5.8) (Pattison and Weber, 2005). 'Polka' and 'Jaclyn' are very susceptible to damage from potato leaf hopper (Plate 5.9), and aphids remain a pest on most varieties, transmitting the mosaic virus complex (Plates 5.10 and 5.11) and other viruses. This is the case in spite of the availability of resistance to most of these diseases in less desirable varieties or in wild populations (Daubeny, 1996).

Resistance to the North American strain of RBDV is conferred by a single dominant gene, *Bu*, in the variety 'Newburgh' and many varieties derived from it, including 'Willamette', 'Chilcotin', 'Malling Promise', 'Cowichan' and 'Nootka'. Additionally, the variety 'Haida' has been shown to be resistant to the resistance breaking strain of RBDV (Kempler and Daubeny, 2008). Numerous genes for resistance to aphids have been identified in *R. idaeus*, *R. occidentalis* and *R. coreanus* (Daubeny, 1996) and probably exist in many other raspberry species. These confer some protection to the mosaic virus complex and other viruses that aphids transmit. Genes for resistance to arabis mosaic, raspberry ringspot and tomato black ring viruses have also been identified in *R. idaeus* (Daubeny, 1996). Resistance to Phytophthora root rot caused by *P. fragariae* var. *rubi* is well known in many varieties, including 'Latham', 'Prelude', 'Anne', 'Nova', 'Josephine', 'Boyne', 'Caroline' and 'Killarney' (Pattison and Weber, 2005). This resistance is polygenic in nature and can be difficult to select due to a large environmental effect, but tends to be persistent over time so that resistance is not lost due to the changing pathogen as is observed in other *Phytophthora* disease, such as late blight in potato.

Development of disease and insect resistance is somewhat more important in black raspberry breeding because one of the major limitations to production is rapid decline of plantings. Mosaic virus, black raspberry latent virus and others cause reduced vigor, poor yields and poor fruit quality and have made the expansion of the black raspberry industry difficult. Verticillium wilt caused by *V. albo-atrum* is the primary root rot disease in black raspberry and is especially prevalent in years with a wet spring followed by high summer temperatures. In such a situation, canes leaf out and then collapse as temperatures rise. Newly emerged primocanes exhibit a characteristic flagging and a bluish tinge (Plate 5.12). Recent screening of wild black raspberry populations from across the natural range of the species has led to the discovery of aphid resistance and *Verticillium* resistance in numerous accessions (Dosset and Finn, 2010). This development with the added emphasis on breeding at the USDA-ARS/Oregon State University program and at Cornell University bodes well for the development of new varieties in the near future.

Adaptation to variable climatic zones, historically, has focused on floricane varieties, especially for expanding the production areas to especially cold regions in the USA, Canada and northern Europe and to hot climates in the south-eastern USA. Recent developments in off season production practices, especially the adoption of high tunnels for raspberry production in California, Mexico, southern Europe and northern Africa has led to the greater emphasis on adaptation of primocane varieties to low chill climates.

Floricane raspberry varieties require a dormant chilling period prior to flowering and fruit production. This period generally occurs in the autumn in temperate climates and is followed by a long winter dormant period where extreme low temperatures can occur, especially at the northern edge of the normal production range. Cold hardiness to temperatures of −30°C (−22°F) or lower is required in these varieties for reliable production. Varieties developed in the USA and Canada east of the Rocky Mountains have traditionally been the most likely to perform well. 'Killarney' and 'Boyne', developed in Manitoba, 'Nova' and K81-6 from Nova Scotia, 'Prelude' and 'Taylor' from New York, and 'Latham' from Minnesota all exhibit strong cold hardiness.

At the opposite extreme of the range for raspberry production in North Carolina and other south-eastern USA states, fluctuating winter and spring temperatures and summer heat can be more problematic than low temperatures. In early spring, temperatures can drastically vary from day to night and from week to week so that many northern varieties begin to break dormancy too early and are susceptible to frost and cold damage. Varieties with a higher heat requirement in the spring are needed so they stay dormant until temperatures stabilize. 'Reveille' and 'Latham' are recommended northern varieties for the south-eastern USA. Further, high summer temperatures are generally not conducive to raspberry production. Optimal temperatures for raspberries are below 25°C (77°F) and photosynthesis quickly decreases as temperatures rise (Fernandez and Pritts, 1994). Breeding for heat tolerance has been done using the subtropical species *R. parvifolius* to develop the varieties 'Mandarin', 'Southland' and 'Dormanred' for regions with fluctuating winter temperatures and high summer temperatures and humidity (Hall *et al.*, 2009).

Additional traits that are more subtle than those discussed so far can also be important in new varieties, such as suckering ability, overall vigor, plant architecture, spine-free canes and the number of fruit. Wild-type plants often produce a large number of new canes each spring that are thin and spiny. Older commercial varieties, such as 'Killarney' and 'Latham', produce an excess of canes and must be pruned aggressively in order to maintain a controlled planting and to enhance pest control measures. Some newer varieties, such as 'Anne' and 'Autumn Britten', are too sparse for traditional perennial production and must be planted in higher densities or in shorter-term plantings where suckering is not as important. Selection for this trait can be done successfully in breeding populations, but the ideal number of canes varies considerably between intended production systems. Off season, short duration plantings in California, Mexico and southern Europe require a few very vigorous canes to produce in narrow rows just months after planting, as in varieties like 'Crimson Giant', 'Maravilla' and 'Isabel'. Perennial plantings require much higher numbers of vigorous canes year after year for a decade or more, thus a more vigorous variety is desirable, such as 'Encore', K81-6, 'Caroline' and 'Heritage'.

Spine-free canes (Plate 5.13) are probably more desirable to the home grower and pick-your-own operator than to large commercial producers. Raspberry spines are relatively small and do not cause considerable problems with professional pickers. However, for the small producer, smooth canes can be an important selling point. Varieties such as 'Joan Squire', 'Joan J', 'Glen Ample' and many others originating in the UK and New Zealand are genetically spineless, carrying the recessive *ss* genotype. New Zealand breeding has reportedly developed spineless black raspberry types with *ss* spinelessness derived from red raspberry, through a multi-generational crossing and back-crossing process begun in 1966 in Scotland by Derek Jennings, but only one, 'Ebony', has been released for the home garden market and cultivars for commercial production are yet to reach the market. The variety 'Canby' is considered functionally spine-free and is often sold as such because it only has spines at the base of the canes. However, this variety is not a good source for breeding spineless varieties because its offspring will contain spines. Many of the cultivars produced in the British Columbian program also are functionally spine-free, and this trait appears to be selected in many of their cultivars.

Growth habit and plant architecture are also commonly selected for in most breeding programs. An upright cane with an open fruiting habit is desired where the fruit are displayed so they can be easily picked. Very few varieties actually meet this standard. 'Meeker', although it is primarily grown for machine harvest and for processing, is such a cultivar from amongst floricane-fruiting types and 'Heritage' is one of the most open fruiting primocane types, especially when grown under tunnels. The canes become somewhat elongated and are thick and upright so that little trellising is needed and the fruit is displayed openly. 'Caroline', on the other hand, is the opposite and produces very large leaves that cover the short fruiting laterals, making finding the fruit challenging. Among floricane varieties, 'Prelude' displays its fruit very openly on upright canes, more so when grown outdoors. Many varieties are overly vigorous under protected production systems so that the fruit tends to pull the fruit into a weeping configuration. The DSA cultivar 'Maravilla' is particularly vigorous when grown in a protected environment, and fruit may be gathered from trellised plants growing 3 m high or more. Trellising is the key for production of this cultivar, and careful training and pruning is essential to keep the fruit visible, making for efficient picking. Stronger, more upright canes are being selected for this fruiting environment in most breeding programs using varieties like 'Heritage', 'Autumn Britten' and 'Polka'.

Raspberry producers and breeders have done a good job in transforming the wild type raspberry into today's important varieties. Improvements can often seem slow in coming due to the time it takes to make and evaluate new hybrid populations and to test potential new varieties. However, compared to many crops, such as apples and grapes, raspberry variety improvement is moving at lightning speed. Tables 5.1 and 5.2 are summaries of many of the most common commercial varieties as well as many new varieties that are just hitting the market and may be available to producers in the near future. The progress made over the last two decades has been significant and with the increasing consumer demand for fresh raspberries the next two decades look very promising for the development of new raspberry varieties. More complete reviews on breeding raspberries can be found in *Raspberry and Blackberries: Their Breeding, Diseases and Growth* (Jennings, 1988), *Brambles* (Daubeny, 1996) and *Raspberry Breeding and Genetics* (Hall et al., 2009).

REFERENCES

Boyles, M.J. and Wrolstad, R.E. (1993) Anthocyanin composition of red raspberry juice: influences of cultivar, processing, and environmental factors. *Journal of Food Science* 58, 1135–1141.

Daubeny, H.A. (1996) Brambles. In: Janick, J. and Moore, J.N. (eds) *Fruit Breeding. Volume II: Vine and Small Fruits*. Wiley & Sons, New York, pp. 109–190.

Dossett, M. and Finn, C.E. (2010) Identification of resistance to the large raspberry aphid in black raspberry. *Journal of the American Society for Horticultural Science* 135, 438–444.

Fernandez, G.E. and Pritts, M.P. (1994) Growth, carbon acquisition, and source-sink relationships in 'Titan' red raspberry. *Journal of the American Society for Horticultural Science* 119, 1163–1168.

Finn, C.E., Moore, P.P. and Kempler, C. (2008) Raspberry cultivars: What's new? What's succeeding? Where are the breeding programs headed? *Acta Horticulturae* 777, 33–40.

Garcia-Viguera, C., Zafrilla, P., Romero, F., Abellán P., Artés, F. and Tomás-Barberán, F.A. (1999) Color stability of strawberry jam as affected by cultivar and storage temperature. *Journal of Food Science* 64, 243–247.

Hall, H.K., Hummer, K.E., Jamieson, A.R., Jennings, S.N. and Weber, C.A. (2009) Raspberry breeding and genetics. In: Janick, J. (ed.) *Plant Breeding Reviews: Raspberry Breeding and Genetics*, Volume 32. Wiley, New York, pp. 39–353.

Hedrick, U.P. (1925) *The Small Fruits of New York*. J.B. Lyons Co., Albany, New York.

Jennings, D.L. (1988) *Raspberries and Blackberries: Their Breeding, Diseases and Growth*. Academic Press, New York.

Kempler, C. and Daubeny, H.A. (2008) Red raspberry cultivars and selections from the Pacific Agri-Food Research Center. *Acta Horticulturae* 777, 1–75.

Kim, S.H., Chung, H.G. and Han, J. (2008) Breeding of Korean black raspberry (*Rubus coreanus* Miq.) for high productivity in Korea. *Acta Horticulturae* 777, 141–146.

National Agricultural Statistics Service (NASS) (2005) Noncitrus fruit and nuts. 2004 Preliminary Summary. United States Department of Agriculture. January 2005. Available at: http://usda01.library.cornell.edu/usda/nass/Nonc FruiNu//2000s/2005/NoncFruiNu-01-25-2005.pdf (accessed 11 December 2012).

National Agricultural Statistics Service (NASS) (2011) Noncitrus fruit and nuts. 2010 Preliminary Summary. United States Department of Agriculture. January 2011. Available at: http://usda01.library.cornell.edu/usda/nass/NoncFruiNu//2010s/2011/NoncFruiNu-01-21-2011.pdf (accessed 11 December 2012).

Pattison, J.A. and Weber, C.A. (2005) Evaluation of red raspberry cultivars for resistance to Phytophthora rot root. *Journal of the American Pomological Society* 59, 50–56.

Roach, F.A. (1985) *Cultivated Fruits of Britain*. Basil Blackwell Publisher Ltd, Oxford, UK.

Weber, C.A. (2003) Genetic diversity in black raspberry (*Rubus occidentalis* L.) detected by RAPD markers. *HortScience* 38, 269–272.

Weber, C.A., Perkins-Veazie, P., Moore, P. and Howard, L. (2008) Variability of antioxidant content in raspberry germplasm. *Acta Horticulturae* 777, 493–498.

NURSERY PRODUCTION OF PLANTS

ALISON DOLAN*

The James Hutton Institute, Dundee, UK

INTRODUCTION

Raspberries are a perennial crop that requires a large initial investment, and so it is essential for them to be pest and pathogen free when planted to give a longer productive life. Establishing a plantation based on healthy plants limits the introduction or addition and build-up of diseases in the existing or neighboring soft fruit crops. This is especially true when the plants are grown on ground not previously used for soft fruit, as it is more likely to be free from soil-borne pests and pathogens. Planting infected material may lead to costly replanting and in some instances the ground itself becoming unsuitable for the commercial growing of raspberries.

Since the introduction, in 1930, of inspection and certification of raspberry canes in the UK, regulations have been in place to provide the soft fruit industry with high health planting material. This material must go through a series of tests before it can be sold as certified stock. The plant must first be tested for a range of pests and pathogens. Once the plant has passed all testing requirements the resultant 'mother' plant is maintained within a high containment, glasshouse facility that is inspected regularly under a statutory regime. This plant and subsequent daughters play an essential role as they are the starting, 'nuclear', blocks in the scheme.

Buds from these plants are sent to propagators along with a Plant Passport and Plant Health Declaration. The buds are initiated and then multiplied through *in vitro* propagation and sold the following spring with a Foundation Grade Certification number. This is issued after a visual inspection by the government Plant Health Inspectors. This visual inspection is to ensure trueness to type and that material is free from pests. The plants produced by the propagators (Foundation Grade) are then grown in nursery 'spawn' beds as per regulations.

*Alison.Dolan@hutton.ac.uk

The material is now eligible for Super Elite Grade Certification for up to a maximum of 4 years as long as it has an unbroken history at that grade. During this time any plants lifted and sold or replanted in new propagation beds will be downgraded to Elite for the next 4 years, then to Standard grade for the final 4 years. After this time the plants can still be sold, but no guarantee of health status will be given.

CERTIFICATION SCHEMES

There are currently three sources of regulation for inspection and certification of soft fruit plants in the UK. The Scottish Government produces rigorous testing requirements for use in Scotland, the Department for Environment, Food and Rural Affairs (DEFRA) produces the UK-wide Plant Health Protection Scheme (PHPS), and in Europe there are the guidelines of the European and Mediterranean Plant Protection Organization (EPPO). The Scottish Government requires that the pathogen testing done at The James Hutton Institute conforms to the guidelines of all of these schemes.

The James Hutton Institute is the sole UK establishment producing pathogen-tested *Rubus* nuclear stock to enter the UK Plant Health Certification Scheme. Although voluntary, the certification scheme is extremely beneficial in preventing the spread of harmful pests and diseases and, as such, is used widely. Healthy stocks are maintained by the regular introduction of pathogen-tested nuclear stock material, maintaining the history of certification and limiting the time that stocks can remain eligible for certification (Plate 6.1). The material entering the scheme is comprised of varieties required by industry as well as new varieties from breeding programs.

The first stage of the scheme works as in Fig. 6.1.

The material that leaves the nuclear stock facility does so with a Plant Health Declaration and Plant Passport. The facility is government-registered to issue passports that give information about the material being released, i.e. genus, variety name, number of buds/root in consignment and destination. It does not give any details of the health status of the material. The Plant Health Declaration, however, does give comprehensive information on the tests done and from what pathogens the material is free; it is signed by a government-authorized signatory. It is this document that is accepted as proof of testing when the material leaves the propagators to enter the certification scheme at Foundation grade.

To work effectively the certification scheme must adapt to changing circumstances, such as the accidental introduction of non-indigenous pathogens on imported commercial plants, characterization and identification of new or existing pathogens, which would lead to new testing requirements, and the use of more advanced detection techniques. Incorporation of new validated detection methods is necessary to provide a scheme that gives confidence and can guarantee the quality of the material available.

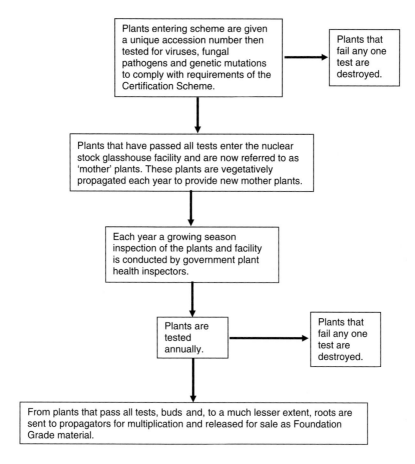

Fig. 6.1. The first stage of the UK Plant Health Certification Scheme.

IN VITRO PROPAGATION

Propagation from raspberry roots was the main method of plant production until the introduction of *in vitro* systems. The advantage of the latter propagation method is that larger numbers of plants are produced in a non-destructive way, i.e. a few buds are harvested from the plant as opposed to the whole root ball as in root propagation. This reduces the number of mother plants required to be maintained, which in turn reduces costs in testing and maintenance of fewer plants. Throughout the growing season short lengths of stem, with buds attached, from the upper part of young canes are taken from the pathogen-tested nuclear stock mother plants. These pieces of cane are supplied to propagators with a Plant Passport and Plant Health Declaration that states the pathogens from which the mother plants have been proven to be free. Once removed from the cane, the buds are surface sterilized and introduced into initiation media followed by multiplication media where, as the name suggests, on average, their numbers double through each transfer into fresh media, a cycle. The number of plants produced is variety dependent. Once multiplied, through no more than ten cycles to prevent genetic changes, they finish in rooting media before they are grown as small plants in aphid-proof glasshouses (Plate 6.2). When hardy, they are planted under Scottish or English/Welsh government-regulated field conditions as Foundation Grade certified stock.

FIELD PROPAGATION

The nursery beds, which must be government approved, are isolated from wild and commercially grown raspberry crops with only autumn-fruiting varieties being permitted to flower because the canes are cut down each year. The beds and land directly next to them must be free from strawberry red core or raspberry root rot disease and all stocks must be labeled with variety and age of plant. Each variety must be free from rogue plants and documentary evidence of soil testing and examination for freedom from potato cyst, dagger and needle nematodes must also be provided. The three grades in the scheme have different requirements, but all the stocks must be free from raspberry root rot, soil-borne virus, bushy dwarf virus and downy mildew. There are two growing season inspections by government Plant Health Inspectors, and depending on their present grade and their freedom from pest and diseases, they are graded as Super Elite (the highest grade), Elite or Standard. A time limit of 4 years is set for each nursery bed and no fungicide treatments for raspberry root rot can be used during this time.

PLANT HEALTH TESTING

It is important that the soft fruit industry has access to planting material that is demonstrated to be free from diseases and genetic disorders. The most common approach is to establish 'mother plants' of each cultivar that are not only tested by a wide range of methods to ensure that they remain pathogen-free, but that they also conform to the phenotype of that cultivar, i.e. retain 'trueness to type'.

Such tests involve visual assessments for virus symptoms on foliage and fruit; bioassays that use herbaceous virus-indicator test plants; graft inoculation to the *Rubus* virus indicator *R. occidentalis* and enzyme-linked immunosorbent assay (ELISA) for raspberry bushy dwarf virus (RBDV). These four techniques are invaluable for the detection of a wide range of viruses.

The nuclear stock mother plants are also tested using molecular diagnostic techniques for *Phytophthora rubi* and *P. idaei*, the oomycetes associated with raspberry root rot disease. All tests described below adhere to the UK scheme and European guidelines, but schemes in other countries will differ in some respect because of the predominance and importance of local pathogens.

TRUENESS TO TYPE TESTING

To ensure that the fruiting characteristics of these vegetatively grown mother plants are not altered by the occurrence of random mutations, which causes 'crumbly fruit' with reduced cohesion of drupelets (Plate 6.3), each year canes of these stock plants are allowed to flower in early spring. The resultant fruits are inspected for aberrations and mother plants that produce abnormal fruits are discarded. In this way, clones of varieties that are prone to develop a genetic mutation are eliminated before being propagated for release to growers.

This work is done in a heated glasshouse separate from the nuclear stock facility and the flowers produced are pollinated by bees from a commercially supplied beehive. The bees are contained and have never been exposed to the environment. They are also fed on an artificial food source so are pest and pathogen free.

VIRUSES

There are a large number of viruses that can infect raspberries (Jones, 1986), many of which have not yet been fully characterized or tests devised. This list of viruses is reviewed as new identification and detection methods are developed and validated for inclusion by the regulatory organizations.

The nuclear stock mother plants are tested for a range of viruses using the UK and EPPO recognized detection methods. The plants are tested during the early part of the growing season, but any graft inoculations made later in the summer are held until the following spring to be assessed. The leaf symptoms displayed on the indicators are best seen on young new growth.

Raspberry bushy dwarf virus is the only pollen-borne virus to infect *Rubus* in the UK. However, it occurs naturally worldwide and can cause symptoms such as yellows disease and/or crumbly fruit. The virus infects the pollen and mother plant and can be readily detected in infected plants by an ELISA.

Raspberry veinbanding mosaic, raspberry yellows, *Rubus* stunt diseases, raspberry leaf curl, raspberry leaf mottle, raspberry vein chlorosis and raspberry yellow spot viruses can all be detected by symptoms.

The following raspberry viruses are detectable by graft inoculation to the indicator *Rubus occidentalis* (Plate 6.4):

- raspberry leaf curl virus;
- black raspberry necrosis virus;
- raspberry leaf mottle virus;
- raspberry yellow spot virus;
- *Rubus* yellow net virus; and
- thimbleberry ringspot virus.

Raspberry viruses detectable by mechanical inoculation of sap to test plants are listed below. The sap indicator plants used are *Cucumis sativus*, *Nicotiana clevelandii* and *Chenopodium quinoa* (Plate 6.5).

- apple mosaic virus;
- *Arabis* mosaic virus;
- bramble yellow mosaic virus;
- cherry leaf roll virus;
- cherry rasp leaf virus;
- cucumber mosaic virus;
- raspberry bushy dwarf virus;
- raspberry ringspot virus;
- *Rubus* Chinese seed-borne virus;
- strawberry latent ringspot virus;
- tobacco ringspot virus;
- tobacco streak virus;
- tomato black ring virus;
- tomato ringspot virus; and
- wineberry latent virus.

Reasons for using biological assays include:

- high sensitivity;
- efficiency, because they detect many viruses simultaneously;
- lower cost, as there is no need for expensive diagnostic materials; and
- some of the viruses have not yet been characterized and no molecular methods exist.

ROOT DISEASES

Raspberry root rot is a serious soil-borne disease caused by the infection of raspberry roots, crowns and bases of young and old canes by oomycetes in the genus *Phytophthora*. Several *Phytophthora* species are known to infect red raspberries, but the most prevalent pathogen in serious disease outbreaks is *P. rubi*, which is very closely related to *P. fragariae*, the cause of strawberry red core disease.

Root rot is now the most destructive disease of raspberries in the UK (Plate 6.6). All parts of the plant below or at ground level can be infected. Affected canes die in the first year of growth or their buds fail to emerge at the start of the second growing season. Alternatively, emerged laterals wilt and die at any time from emergence through late in the fruiting season.

Although *P. rubi* is generally considered the most important pathogen causing raspberry root rot, other related pathogens such as *P. idaei*, *P. cambivora*, *P. citricola*, *P. cryptogea* and *P. megasperma* have also been found to infect red raspberry roots (Duncan *et al.*, 1987).

Recent work looking at the potential threat to the Scottish raspberry industry from *P. idaei* has shown that although this pathogen does damage the roots of infected plants it does not appear to have a significant effect on yield (Dolan and Cooke, SCRI, unpublished data).

In the field, *Phytophthora* species produce resting oospores that can remain viable in the soil for up to 20 years. Thus, replanting raspberries into contaminated soil can result in the germination of oospores and the release of motile zoospores that are readily spread in soil water and infect the new crop. It is for this reason that nursery beds are grown on land void of this disease and that fungicide treatments against it are not permitted as they only control, not eradicate, the pathogen.

In the glasshouse, before entry into the nuclear stock facility, two samples of root from the candidate material are tested by a polymerase chain reaction (PCR) technique for *P. rubi* and *P. idaei*. These samples are taken mainly from the base of the root ball and from any apparently unhealthy brown roots, i.e. areas where infection will be detected if present. The two samples are taken independently from different parts of the ball so they are, without taking too much root and killing the mother plant, representative of the whole root ball. Despite this and in case small areas of infected root have been missed, further precautions are taken after entry into the nuclear stock glasshouse. It is essential to prevent potential spread of motile zoospores from an individual plant within the population of mother plants and this can be achieved by using a series of preventative measures. Plants are individually irrigated using spray stakes that cannot back siphon and are spaced well apart on weld mesh benches (Plate 6.7). A new pair of disposable gloves is worn when handling each plant. To prevent disease coming into the propagation system from other sources, the mother plants are grown in compost that has been sterilized by autoclaving, irrigated and fed using water that has passed through a UV sterilization system. Access to the facility is restricted to essential personnel wearing coveralls and shoe covers. The mother plants are then tested each year for *P. rubi* and *P. idaei* so that only healthy roots are grown to provide the following year's mother plants.

These procedures also reduce the risk of other soil-borne fungal and bacterial diseases, such as Verticillium wilt and crown gall.

Nursery beds are established where raspberries have not been previously grown, and therefore, where raspberry root rot should not be present. If any plants are suspected of having the disease, those and the surrounding plants will be tested. If disease is confirmed then all infected plants will be destroyed. This area of land will no longer be eligible for nursery plant production for certified stock.

It is also important to ensure that biotrophic fungal pathogens are excluded from the mother plants by regular inspections. For example, downy mildew (*Peronospora sparsa*) particularly affects blackberries and blackberry × red raspberry hybrids, leading to systemic infection that can be difficult to detect without regular inspection of foliage and microscopic examination of putative lesions, or the use of a molecular test. Another biotroph of particular concern during propagation is orange rust (*Arthuriomyces peckianus*), because it generates systemic mycelium that spreads into the root system.

Verticillium wilt is caused by two common soil-borne fungi, *Verticillium albo-atrum* Reinke & Berthier and *V. dahliae* Kleb. The former is reported to be more prevalent in humid climates and the latter more common under semi-arid conditions. In red raspberries, the disease is less severe than in black raspberries and blackberries. Infected plants can survive for years, albeit with reduced vigor, fewer primocanes and smaller fruit. The disease is not considered a serious problem in the UK or across Europe in open field production, and as such, is not tested for under any of the schemes at present. However, this may change in the future as much of the *Rubus* industry in Europe and North America is now grown out of season under plastic tunnels or in glasshouses where temperatures are higher and the risk of Verticillium wilt greater.

CANE DISEASES

Although not observed in the glasshouses where the nuclear mother plants are grown, the following cane diseases may be found in the nursery beds, although they are rarely considered to be a problem.

Cane Botrytis caused by the fungus *Botrytis cinerea* Pers: Fr. is an important cane disease. The fungus prefers mature tissue, producing lesions by infection from airborne spores on the older leaves of primocanes. It is a wound pathogen infecting tissue damaged by machine, insects, frost or herbicides. The lesion develops throughout the leaf and down the petiole. This can cause bud failure and in severe cases can completely girdle and kill the primocane. Black sclerotia protect the hyphae over winter on canes and the disease cycle begins again with the release of spores in the spring during wet weather (Plate 6.8). The disease is found where there is a dense canopy, so to reduce its severity it is advisable to remove excess growth between rows and limit the numbers of canes tied in to maintain a free flow of air and good light conditions. Nursery beds are usually grown at high density and cane Botrytis can develop even in the first season. However, most of the over-wintered canes that are used to establish new field crops are cut almost to soil level in spring after planting, thus removing the inoculum. Fruits found on autumn-fruiting varieties may also become infected with grey mold.

Spur blight, caused by the fungus *Didymella applanata* (Niessl.) Sacc., is often found infecting the same canes as cane Botrytis. The main infection risk is from splash-dispersed pycnidiospores infecting mature leaves of primocanes. Infection causes stunted growth of axillary buds, sometimes resulting in failure to produce lateral buds in the following season, subsequently leading to some yield loss. Cane Botrytis is thought to be more damaging than spur blight.

Cane blight is caused by the fungus *Kalmusia coniothyrium* (Fuckel) Huhndorf (previously named *Leptosphaeria coniothyrium* (Fuckel) Sacc.) and *Coniothyrium fuckelii* Sacc., which is another wound pathogen. The fungus infects primocanes wounded in a variety of ways, such as by summer pruning, machine harvesting or by abrasion of primocanes against old cane stubs. The vascular lesion can cause the laterals to wilt or in severe cases kill canes over winter or into the following spring by cutting off the supply of food and water to the upper part of the plant. The lesions can be seen clearly during winter by scraping to remove the primary cortex and exposing the brown vascular tissues, which should be green if healthy. Cane vigor control and biennial cropping resulting in a reduction or elimination of the growth of the primocanes during harvest helps reduce the infection, and thus the severity of the disease.

Cane spot is caused by the fungus *Elsinoe veneta* (Burkholder) Jenk. Although mostly found on canes, it can also be seen on leaves, petioles, pedicels, flower buds and fruit. The fungus infects young green tissue with later infections in the growing season, but not as severely. The infection causes purple spots to appear, mainly on the middle area of the stem, which progress into sunken pits. These pit lesions can merge, forming deep cankers that can kill the cane. To prevent further dispersal of the fungus, overhead irrigation should be removed, and as with many of the other canes diseases the plants should be maintained in a less humid environment, avoiding a dense canopy by cutting out excess canes and removing weeds.

LEAF DISEASES

The symptoms of raspberry leaf and bud mite are caused by a microscopic eriophyoid mite. It is impossible to see with the naked eye and is often missed until symptoms in the plant become apparent. The females overwinter under bud scales and in petiole scars. When the new growth appears, the mites emerge and migrate to new shoots and leaves. If very high populations are present during fruiting, the berries become infested. The symptoms of infection are very dependent on the age of plant at the time of infestation and the environmental conditions (Plate 6.9). Infestations are most severe in sheltered conditions.

Raspberry yellow rust is caused by the fungus *Phragmidium rubi-idaei*. In early spring, it can infect the upper surface of young leaves, producing yellow aecia. As the summer progresses yellow uredinia can be found on the underside of the leaves. These pustules turn black later in autumn as overwintering teliospores develop. Premature loss of leaves can cause some loss of winter hardiness or reduction in cane vigor in the following season. Inoculum levels can be reduced by using a biennial cropping system or removing old cane stubs and leaf litter. Teliospores adhering to old cane stubs and overwintering canes are the sole survival method for this fungus. If this or any other rust is observed during statutory field inspections of nursery plants then the plants are downgraded.

Powdery mildew is caused by the fungus, *Sphaerotheca macularis*, and can infect leaves as well as young and mature fruit. It thrives under warm, dry conditions, and so varieties cultivated under tunnels can be particularly susceptible. All nursery beds are grown in open fields with very few fruit left to develop, so this fungus causes little problem to the crop.

ADDITIONAL INFORMATION

Enquiries about the Certification Scheme in Scotland should be directed to: hort.marketing@scotland.gsi.gov.uk.

Additional information can be found at:

- http://www.fera.defra.gov.uk/plants/plantHealth/phps.cfm
- http://www.eppo.org/STANDARDS/standards.htm
- http://www.fruithealth.co.uk/

REFERENCES

Duncan, J.M., Kennedy, D.M. and Seemüller, E. (1987) Identities and pathogenicities of *Phytophthora* spp. causing root rot of red raspberry. *Plant Pathology* 36, 276–289.

Jones, A.T. (1986) Advances in the study, detection and control of viruses and virus diseases of *Rubus*, with particular reference to the United Kingdom. *Crop Research* 26, 127–171.

Propagation

Courtney Weber*

Department of Horticulture, Cornell University–New York State Agricultural Experiment Station, New York, USA

INTRODUCTION

Although raspberry plants easily multiply naturally, it is wiser for a grower to purchase disease-free nursery stock rather than to dig canes from their own or their neighbors' plots. For newer patented varieties, it is also illegal to do this without obtaining a license and paying the royalty fee to the patent holder. Regardless of the legalities, the risk of spreading disease through contaminated soil particles and infected plant tissue is significant and can mean the difference between the success and failure of a planting. Certified virus-indexed plants grown in fumigated ground or soil-less growing media are the best option for growers, will have the best growth, be the most productive and live the longest (Funt *et al.*, 1999).

Nursery stock is commonly available as bare root plants (dormant suckers, matured transplants, or tip-layered canes (black and purple raspberry)), tissue-cultured plants (plugs or nursery matured) and bulk root pieces. However, not all nurseries produce all types of stock.

The perennial crown and suckering nature of the red raspberry provides the basis for traditional commercial propagation of bare root plants and bulk root pieces. New canes (suckers) are produced from crown buds or from adventitious root buds, especially in the spring after winter chilling. Suckers from adventitious root buds are the traditional transplant type for open field perennial production of red raspberries. These can be dug as individual canes for use in a commercial field or replanted in the nursery to grow for another year to produce a larger plant. The year-old sucker transplants are more vigorous and establish more readily, but are more expensive than newly dug suckers so they are not generally used for commercial plantings (Funt *et al.*, 1999).

---

* caw34@cornell.edu

Root cuttings can also be used, either to produce red raspberry suckers in the nursery or to directly establish plantations for fruit production. The large root system of the raspberry plant grows much more extensively than what is required for a healthy bare root plant. The excess root mass can be used for propagation and can be purchased as bulk roots from some nurseries.

Black and purple raspberry canes mostly grow from basal buds in a strong crown and do not produce many suckers from the root system, so field propagation is primarily done by tip layering. This is the primary field propagation method for black raspberries and some purple raspberries, but is less necessary in purple varieties with a greater percentage of red raspberry in the pedigree, such as in 'Royalty', which suckers more like a red raspberry variety. Tip layering is usually done in late summer by burying the tips of the current season's cane 5–10 cm (2–4 in) in the soil. The buried tips develop roots and form new plants before dormancy in the same year. These are cut from the mother plant and dug in the dormant season for sale as bare root plants.

Additionally, the use of tissue culture for mass-producing raspberry plants has become common for commercial propagation. Plantlets are produced for direct rooting into plugs (green or dormant) for production field use or to be planted in the nursery for a year-long maturation period (nursery matured stock). Tissue culture plants consistently have fewer diseases and, if established well, are more productive than other sources of plants.

TISSUE CULTURE

In vitro tip or tissue culture techniques for propagation of all raspberry types were developed in the late 1970s and early 1980s (Kiss and Zatykom, 1978; Anderson, 1980; Pyott and Converse, 1981) and are routinely used by a number of nurseries. When derived from virus-free parent plants and rooted in sterile potting media, this type of transplant should be free of most diseases, insects and nematodes. Tissue culture transplants may be acquired in several forms:

- still in culture vessels (or recently removed from culture vessels);
- established in rooting cubes as either green or dormant plugs; or
- bare-root, dormant plants (nursery matured) that have grown for one season in nursery rows in fumigated field soil.

Plants established as plugs are the best choice for establishing a new, disease-free field with tissue culture derived plants, because the plants can be planted directly in the field after the danger of spring frost has passed. Using tissue culture plugs avoids insect, disease or nematode contaminations, which can occur even on fumigated nursery land. Nursery-matured plants may reach maturity sooner, and the larger root system may allow survival under drier soil conditions during the planting year. This is especially true for black raspberries because the tissue culture plugs are very sensitive to herbicides and drought as they develop the crown needed for perennial growth. Red raspberries are less sensitive because they develop a strong root system from which the canes will emerge, often during the first season.

New cultures of raspberries can be initiated from dormant axillary buds, leaf discs or other tissues (McNicol and Graham, 1990). While many of these techniques and tissues have been successfully used to initiate new cultures, axillary buds from primocanes harvested in mid-summer are the preferred material.

The meristematic tips of elongated axillary buds that have been forced in sterile culture are harvested and induced to multiply. To initiate, canes are cut into single nodal segments, leaving 2–3 cm (~ 1 in) on each side of the axillary bud (Plate 7.1). The segments are surface sterilized using 70% ethanol and sodium hypochlorite (common household bleach). Several different pre-sterilization treatments can be used to reduce the contamination level in the new cultures, but the most effective is growing the source plants in a greenhouse, which reduces fungal and bacterial contamination. Rinsing under running water for 10–20 min, a 10-s agitation in a 10% solution of household disinfectant, or a 10-s agitation in 250 ml of water with five drops of grease-cutting liquid dishwashing soap have all shown some effectiveness in reducing fungal contamination. This is followed by a 30-s soak in 70% ethanol then 4–5 min in 10% bleach (0.6% sodium hypochlorite) with a few drops of surfactant per liter of solution (Tween 20, Triton-X, liquid dish soap or others). The bleach is rinsed off three times with sterilized water. A final dip in 70% ethanol helps to wick away the water. The nodes can then be placed in a sterile covered Petri dish or on another work surface to allow the ethanol to evaporate. Sterilization and subsequent preparation of shoot segments must take place in a laminar flow hood or other sterile environment to avoid recontamination of the segments.

To start the culture, the nodes are trimmed to remove the cut edges that were exposed to the sterilization solutions (Plate 7.2). The nodes are then placed upright in the nutrient culture medium to allow the axillary buds to elongate (Plate 7.3). Glass test tubes, GA7 boxes, glass baby food jars and/or plastic deli tubs can be used as the culture container. Many culture medium recipes have been developed for various tissue culture purposes. A detailed discussion of the importance of different nutrients and their sources can be found in In Vitro Culture of Higher Plants (Pierik, 1987) and Plant Propagation Principles and Practices (Hartmann and Kester, 1983) as well as many other sources.

For raspberries, a basic Murashige and Skoog (MS) medium (Murashige and Skoog, 1962) or some derivation of this medium is most commonly used to initiate and grow new cultures. Modifications including reduced macronutrients (Anderson, 1980), Knudson's salts (Knudson, 1946) and others (Pyott and Converse, 1981) have been used in the growth medium.

Varying hormone strengths, combinations and timing have also been utilized (Hoepfner, 1989). In most cases, a basic MS medium with 1 mg 6-benzylaminopurine/benzyladenine (BA) per liter of medium can be used to initiate new cultures. Various modifications have been made to the basic medium to improve the growth of the cultures, including the addition of 100 mg/l myo-inositol and removal of the iron from the MS micronutrient mixture and replacing it with 270 mg/l of chelated iron derived from sodium ferric ethylenediamine di-(o-hydroxyphenphenylacetate) found in commercial iron fertilizer supplements, such as Sprint® 138 (Becker Underwood, Inc., Ames, Iowa) rather than iron chelated with disodium salt of ethylenediaminetetraacetic acid (NA_2EDTA) which is commonly used. The addition of 0.9 ml/l of a 1:10 dilution of a fungicide containing iprodione as the active ingredient (formulation containing 477.5 g/l (4 lb/gal) of iprodione) to the medium is useful in reducing fungal contamination.

The bud will begin elongation in a few days and can be harvested after 10–20 days (Plate 7.4). If possible, the bud should be left to elongate for 30 days or longer before excision, which encourages multiplication to begin. These shoots are most easily rooted and/or multiplied. The excised shoot is placed in medium without hormones initially for 2–3 weeks to stabilize the growth. The addition of BA and indole-3-butyric acid (IBA) can be used once the culture is stable and growing to induce multiplication. Axillary buds that root in the absence of hormones can be induced to produce more shoots by tipping the plantlets and leaving the tipped plant to grow, thus forcing additional axillary buds to elongate. The harvested tip can be used to initiate a new culture.

Rooting can be done *in vitro* or *ex vitro* in soil-less medium. The addition of IBA (0.5–1.0 mg/l) or 1-naphthaleneacetic acid (NAA) (0.18 mg/l) to the tissue culture medium will induce rooting, although often the roots are somewhat abnormal. This does not seem to be problematic to establishing plants in soil. Alternatively, cultures can be etiolated by reducing or eliminating light for 3–5 days before attempting rooting. These etiolated shoots will root without hormone *in vitro* or in a soil-less potting medium under mist. The use of a 0.1% IBA rooting powder will aid rooting in the soil-less medium. Optimization of procedures and medium for each variety is often necessary and usually beneficial in mass production of tissue culture plants.

Rooting tissue culture plantlets *ex vitro* is ideally done in soil-less medium under mist or under humidity domes. Soil-less potting mixes commonly contain some combination of peat moss, perlite, vermiculite, sand and pine bark. Dolomitic limestone and various fertilizer mixtures can also be added to improve growth. Avoid using soil for propagation because of the risk of disease contamination and rules against transporting soil to different parts of the USA and world.

Tissue culture plugs are sold as green plants, dormant plugs or planted in fumigated nursery ground for a year of growth and maturation. Green plugs are rooted during the winter and spring for spring and/or fall planting, depending on the production location. Dormant plugs are rooted in the summer and allowed to go dormant in the fall. They are shipped dormant for spring planting. Nurseries that produce nursery-matured tissue culture plants establish plantings in the spring and dig the following dormant season. During this maturation period, the plants expand their root system but sucker lightly or not at all and are handled in a similar way to bare root suckers in the following planting season.

DORMANT PLANTS

Dormant plants are the most versatile plant type for starting new plantings and are often the most widely available and convenient choice for growers. Daughter plants, field transplants and nursery-matured tissue culture transplants are mechanically dug and shaken to remove soil during the dormant season for sorting, storing and shipping. Plants dug and held dormant in storage, if properly handled, are usually as good as freshly dug ones. Once dormancy is broken, the utility of plants for establishing new plantings quickly declines due to the risk of transplant shock.

Red raspberry suckers, matured transplants and nursery-matured tissue culture plants should retain a portion of the parent plant root so that they can take up water and establish well. They may have an inverted 'T' or an 'L' shape with about 14 cm (~ 5.5 in) of the cane (called the 'handle') remaining (Funt et al., 1999) for ease of planting. Rooted black raspberry tips are also dug while fully dormant. Before digging, they are cut from the original plant, also leaving a handle.

The handle is part of an overwintered floricane and will push flowering laterals and eventually die during the first season. Flowers should ideally be stripped from the handle of new transplants to encourage root and sucker growth. Transplants with relatively large root systems are acceptable, but those with few or no roots and no part of the parent root are not likely to survive.

Excess red raspberry roots can be cut and separated from the dormant canes for sale as bulk root cuttings, to establish new nursery beds or to be discarded. Roots of variable lengths and 1.5–4.5 mm ($^1/_{16}$–$^3/_{16}$ in) in diameter are most desirable for starting new plantings and should be placed at about a 7.5 cm (3 in) depth in the soil with approximately 50–60 g (2 oz) of roots per hill or per meter (~ 3 ft) of trench.

Long canes are a special class of dormant canes dug for off-season production in the late winter, usually in areas with a Mediterranean climate like southern Spain, Portugal, northern Africa and southern California. This type of stock is produced by growing the plants in a cool, temperate climate where plants can go dormant naturally and are dug before the ground freezes.

The canes are left intact or trimmed to 1.5–2 m (5–6.5 ft) in height and must be carefully dug to avoid damaging the fruiting wood. The plants are chilled artificially for 6–8 weeks prior to planting for late winter production. The chilling requirement varies widely among varieties with true floricane types, such as 'Glen Ample' and 'Tulameen', requiring more than those that set a few primocane flowers, such as 'Glen Lyon'. Digging varieties with a strong floricane habit too early in the season, before flower bud development, can lead to lateral buds breaking without flowers and little to no production. Long canes are very expensive and mortality may be high due to the high water stress on a limited root system or if roots dry out during storage. However, fruit prices are very high during late winter, making the system attractive to some growers.

For growers in colder climates, a modified system of growing for greenhouse production in late winter has also been developed (Koester and Pritts, 2003). In this system, plants are grown outdoors in pots and either left out to chill naturally or moved into coolers in the fall after becoming dormant for artificial chilling. The potted plants are then moved into greenhouses in early winter for forcing and late winter/early spring production. This system is expensive and labor intensive but has been shown to be feasible for growers with available greenhouse space prior to the spring bedding plant season.

For the homeowner, newly emerged red raspberry suckers (primocanes) may be transplanted in early spring when suckers are 12–20 cm (5–8 in) tall. However, care must be taken to provide adequate moisture and weed control. This is not commonly done on a commercial scale due to the added care and increased mortality of transplanted canes.

STORAGE AND HANDLING

In the case of plants produced for bare root sales, once the canes are dug, the soil is removed and the canes allowed to briefly dry. Storage of bare root canes in the dark at –2°C to 0°C (28–32°F) is recommended. Roots must be wrapped or otherwise covered to keep them from drying out, while also limiting the moisture level to prevent rotting and ice formation. Slightly moist sawdust can be used as an insulator to keep the plants healthy. Storing the canes with any source of ethylene such as ripe fruit, especially apples, should be avoided as this can cause damage to the canes.

Tissue culture plugs can be started in the summer and allowed to go dormant naturally in the fall in cold climates when hardened off and placed outside or in unheated greenhouses. Chilling can be achieved by storing the plants at 7–10°C (45–50°F) for 8–10 weeks. The plants can then be stored at 2°C (35°F) until shipping, ensuring that the root ball does not dry out during storage. Green plug plants can be rooted so that they are timed to be ready for planting at the optimal planting time for a given growing region.

SHIPPING

Plants should be shipped as rapidly as feasible. Bare root plants should be wrapped or packed with moist sawdust, burlap or shredded paper to avoid drying out during shipping. They should be placed in a thick grade plastic bag that will resist tears and punctures. If shipped in trays, dormant and green plugs must be packaged securely to ensure the plants remain within their plug tray wells. Many approaches are used to accomplish this and usually consist of some sort of chambered or stackable cardboard units. Plants are also sometimes shipped after being removed from trays and should be securely packed in boxes.

RECEIVING THE PLANTS

After receiving the dormant plants, the packages should be opened immediately and, if necessary, the roots should be moistened. Unless planting can be done within a few days, the plants should be placed in storage at about $2°C$ ($35°F$), or heeled in until they can be planted. To do that, a shallow trench should be dug, deep enough to accommodate the root systems, in a sheltered and shaded area where the soil is well drained. Bundles should be opened and a single layer of plants placed against one side of the trench so that the root systems are completely below the soil surface. Roots should be covered with soil and firmed carefully. Plants so treated can be held safely for a reasonable length of time (1–2 weeks) if they are not allowed to dry out. Setting plants should not be delayed any longer than absolutely necessary. It is important to keep plants from drying out.

When any freshly dug or tissue culture plant is shipped with living or growing tissues, it requires a 'hardening' process before being put into the field. Plants should be set outdoors in soil, if necessary, and placed in a shaded area for 2–3 days and moved to full sun gradually over a 5–7-day period. Moisture should be maintained with frequent or daily watering and plants should be protected from any hard freezes ($-2°C/28°F$ or lower).

The propagation of high-quality plants to start new raspberry plantations is of the utmost importance and is the foundation of the entire industry. Nurseries follow a basic plan of fumigation, pest management and plant culture that can provide an excellent product to growers. Failure at the nursery level to maintain clean plants, package and ship plants quickly and securely, and maintain varietal integrity means failure at the farm. Most commercial nurseries are happy to work with growers to ensure the product they receive will help them be successful. It is acceptable and often welcome when a grower contacts a nursery to discuss the grower's needs and plans so that the nursery can provide the product needed at the right time and in optimal condition.

REFERENCES

Anderson, W.C. (1980) Tissue culture propagation of red and black raspberries, *Rubus occidentalis* and *R. idaeus*. Acta Horticulturae 7, 13–20.

Funt, R.C., Bartels, S., Bartholomew, H., Ellis, M., Nameth, S.T., Overmyer, R.L., Schneider, H., Twarogowski, W.J. and Williams, R.N. (1999) *Brambles: Production, Management and Marketing*. Bulletin 782. The Ohio State University Extension, Columbus, Ohio.

Hartmann, H.T. and Kester, D.E. (1983) *Plant Propagation Principles and Practices*. Prentice-Hall, Inc., Englewood Cliffs, New Jersey.

Hoepfner, A.S. (1989) *In vitro* propagation of red raspberry. Acta Horticulturae 262, 285–288.

Kiss, F. and Zatykom, J. (1978) Vegetative propagation of *Rubus* species *in vitro*. Botanika Kozlemenyek 65, 65–69.

Knudson, L. (1946) A new nutrient solution for the germination of orchid seed. Bulletin of the American Orchid Society 15, 214–217.

Koester, K. and Pritts, M. (2003) *Greenhouse Raspberry Production Guide*. Department of Horticulture Publication 23. Cornell University, Ithaca, New York. Available at: http://www.fruit.cornell.edu/berry/production/brambleproduction.html (accessed 27 September 2012).

McNicol, R.J. and Graham, J. (1990) *In vitro* regeneration of *Rubus* from leaf and stem segments. Plant Cell, Tissue and Organ Culture 21, 45–50.

Murashige, T. and Skoog, F. (1962) A revised medium for rapid growth and bioassays with tobacco tissue cultures. Physiol Plant 15, 473–497.

Pierik, R.L.M. (1987) In Vitro *Culture of Higher Plants*. Martinus Nijhoff Publishers, Boston, Massachusetts.

Pyott, J.L. and Converse, R.H. (1981) *In vitro* propagation of heat-treated red raspberry clones. HortScience 16, 308–309.

SITE PREPARATION, SOIL MANAGEMENT AND PLANTING

Marvin Pritts[1]* and Eric Hanson[2]

[1]*Department of Horticulture, Cornell University, New York, USA;* [2]*Department of Horticulture, Michigan State University, Michigan, USA*

INTRODUCTION

The treatment and management of the site and soil before planting can impact the growth and productivity of raspberries for years to come. Many consider the year prior to planting to be the most important because of the substantial effects that soil factors have on plant growth. Modifications to soil are much easier to make without raspberries present on the site (Plate 8.1), hence the emphasis on modifications before planting. Furthermore, changes to soil properties take time, and waiting until just before planting to attempt modifications may not be sufficient.

Chemical, biological and physical soil components each need evaluation and may require substantial modification. Even if the plants will perform adequately without modification, it is almost always the case that improvements to soil can be made. These improvements, such as optimizing the soil pH and increasing organic matter content, can take months.

PHYSICAL PROPERTIES

Soil physical properties are difficult to modify, so it is important to select a site with an appropriate soil type. Raspberries grow best in well-drained, loamy soils, although they are fairly tolerant of a range of soil types. Regardless of soil type, they are not very tolerant of the poor drainage often associated with compacted soils or those with a perched water table. It is critical to provide drainage to soils that retain free water in the top 0.5 m (20 in) after rainfall, or plant on raised beds to facilitate drainage (Wilcox *et al.*, 1999).

*mpp3@cornell.edu

© CAB International 2013. *Raspberries* (eds R.C. Funt and H.K. Hall)

The cation exchange capacity (CEC) is made up of contributions from both the mineral and organic components of the soil. In general, a high CEC indicates that the soil is relatively fertile and has the ability to hold onto potassium (K), magnesium (Mg) and calcium (Ca) ions. Although the mineral contribution to CEC cannot be modified significantly, the organic component can be increased through the addition of organic matter, such as manure, compost or incorporated cover crops. Increasing the organic matter generally increases the CEC and the water holding capacity of the soil. It also improves soil structure, the aggregation of soil particles and soil tilth. Increasing organic matter and water-holding capacity are usually always beneficial. In wetter, heavier soils where soil pathogens may be present, increasing water-holding capacity through the addition of compost can be detrimental, but one should not be planting raspberries in such soils.

Increased aggregation of soil particles facilitated by organic matter allows for better soil structure and root development. Without structure, soil particles collapse on each other, resulting in compaction and small pore spaces between particles. Soils without pore space have a difficult time holding onto water; water tends not to drain through the soil profile, and roots have a difficult time growing and penetrating the soil. Soil structure is also damaged when soil is repeatedly cultivated. Although a field that has been rototilled several times may look good prior to planting, the soil structure may be degraded so that when it rains, the soil becomes compacted with little pore space for air and water. Also, repeated tillage and the subsequent exposure of organic matter to oxygen will accelerate decomposition of the organic matter. Driving heavy equipment onto a field when soil is wet also increases compaction. Each of these practices should be avoided. If a field does have a plow layer or is otherwise compacted, then deep subsoiling may be required to break up the compacted zone. Repeated shallow plowing or tilling will only increase compaction and soil degradation.

Green manure cover crops can be incorporated to increase soil organic matter, providing nitrogen (N) if the cover crop is a legume. They also tend to suppress weeds while growing. Obviously cover crops need time to grow, so plan for site preparation and soil modification well in advance. Popular pre-plant cover crops for raspberries include peas, lupines, alfalfa, buckwheat, rye, wheat, oats, clover, Sudan grass, vetch and canola.

BIOLOGICAL PROPERTIES

Living organisms in the soil can have a significant impact on plant growth. These include bacteria, viruses, fungi, nematodes, protozoa, algae, insects and earthworms. The vast majority of these are beneficial or benign. These organisms recycle nutrients and bind soil particles together, improving soil structure. Although the role of specific microorganisms may not be well understood, some generalizations can be made. Soils with a large amount of organic matter to sustain these organisms and fuel biological activity tend to be healthier and more stable than those without much activity. This biological activity tends to suppress harmful organisms that attempt to become established. Organic matter that is 'aged' tends to have a more stable complex of organisms than fresh material. Some composts have been shown to suppress plant pathogens in the soil and reduce plant disease.

Studies have generally shown that incorporated compost either has no effect (Hargreaves et al., 2008) or enhances plant growth (Black et al., 2003; Forge et al., 2009) in raspberries. When pathogen pressure is high and soil drainage is poor, added compost has the potential to reduce plant health. The incorporation of beneficial microorganisms, such as mycorrhizae or bacteria, into soil before planting, or inoculating plant roots prior to planting, has been promoted as a way to enhance growth and productivity. There is only scattered evidence that this is an effective practice (Toussaint et al., 1997; Taylor and Harrier 2000; Orhan et al., 2006), especially under field conditions. Providing a growing media for beneficial organisms (e.g. compost and organic matter) seems to be critical for obtaining consistent suppression of soil pathogens.

Certain cover crops can also help suppress disease-causing organisms in raspberries (Seigies et al., 2006). Marigolds, mustards and certain cultivars of oats and Sudan grass, for example, will suppress nematodes when grown prior to planting raspberries. Testing for harmful nematodes is recommended before planting, particularly because they can harbor and transmit virus diseases. If levels are very high, fumigation may be recommended as opposed to natural means of nematode suppression (Belair, 1991). One risk of fumigation is that it kills all soil organisms, even those that are beneficial and benign. Beneficial organisms may not re-establish well after fumigation, leading to an increased risk of pathogen infection in later years. Solarization offers an alternative to fumigation where light levels are high (Pinkerton et al., 2002). Cover crops are usually a better choice if sufficient time is available to grow them prior to planting. Fumigation will suppress these pathogens quickly, but it is very expensive and difficult to apply properly.

Among the most detrimental organisms present before planting are weeds. Cover crops can help suppress weeds, especially those that grow fast and tall. Suppression is greatest when a broad-spectrum, non-selective, non-residual herbicide is used to kill perennial weeds before cover cropping. Repeated cultivation can also reduce weed pressure, but is detrimental to soil structure.

A sequential mixture of cover crops appears to work best for suppressing a range of harmful organisms while improving organic matter. Many growers use a combination of summer and winter cover crops in sequence. For example, summer cover crops, such as Sudan grass or buckwheat, may be followed with vetch or rye for the fall/winter. Plants will almost always perform better following 1 or 2 years of cover crop rotations prior to planting.

CHEMICAL PROPERTIES

The soil is composed of many chemicals, which can be naturally occurring or added by human activity. For the most part, the soil is a large reservoir of minerals and organic material that slowly releases chemical elements, some of which are taken up and used by plants as essential nutrients. Most evaluation methods seek to estimate the amount of these nutrients available to the plant for growth and development – these are generally grouped into those that are required only in small amounts: boron (B), iron (Fe), zinc (Zn), manganese (Mn), copper, (Cu), molybdenum (Mo) and chlorine (Cl), and those required in larger amounts: nitrogen (N), phosphorus (P), potassium (K), calcium (Ca), magnesium (Mg) and sulfur (S). Soil amendments usually consist of fertilizers containing nutrients from this latter group. Certain other elements may be beneficial for certain plants, but are not essential for growth. These are silicon (Si), cobalt (Co), nickel (Ni), selenium (Se), sodium (Na) and aluminum (Al). The role of these particular nutrients is not well understood in berry crops.

Soil samples for testing should be taken randomly (or systematically to ensure thorough representation) from a field to be planted. A minimum of ten samples should be taken from a field that has the same soil type and cropping history. A sampled section should be 4 ha (10 acres) or less. If the field is larger than 4 ha (10 acres), then a second test should be conducted. The samples should come from the future root zone, combined in a bucket and mixed. A composite sample should be sent to a soil testing laboratory where extraction and analysis will occur. Different labs use different extractants, so the estimate of the amount of essential plant nutrients will differ from lab to lab. Recommendations should be based on the specific extractant used by the lab. Do not use results from one lab to generate recommendations from another.

The first test to consider is the soil pH. The optimal pH for raspberries is 6.0–6.5 (Plate 8.2). If the soil pH is outside this range, then nutrient uptake can be compromised (Plate 8.3) and an amendment will be recommended. Sulfur is used to lower pH and lime is used to raise pH to the target of 6.5. The amount of S or lime required to change pH to the optimum is dependent on the ability of the soil to hold onto alkaline-forming cations such as K, Mg and Ca. Typically, heavier soils and those with high organic matter (a high CEC) require more amendment to change pH than sandier soils with low organic matter (a low CEC). The soil test result should provide a recommendation for the amount of S or lime to add, and will indicate if the lime should have a high proportion of Ca (calcitic) or Mg (dolomitic). A common myth is that certain organic materials, such as pine needles, can lower soil pH. Studies have shown that incorporated organic materials, while perhaps increasing the CEC slightly, have little effect on soil pH.

Lime is not 100% pure, so it will come with a percentage that represents its effective neutralizing value (ENV). For example, if the lime has a 90% ENV rating, then it will take 110% of the recommended rate to bring about the desired change in soil pH. Both lime and S are available in pelletized form to facilitate spreading and to reduce dust and blowing. However, pelletized forms take longer to break down and react with soil to change pH. Most soil testing labs assume a 15 cm (6 in) slice of soil for making recommendations. For raspberries that have a slightly deeper rooting depth (i.e. 20 cm, 8 in), amounts should be increased by 25% (8/6) to bring about a change in pH to a depth of 20 cm (8 in).

The lime or S should be broadcast over the entire area and incorporated to the depth of the rooting zone. Correcting the soil pH with lime will usually provide all of the required Ca and Mg as well. Some soils may require supplemental Mg even when the pH is 6.5 and no lime is required. In these situations, magnesium sulfate can provide additional Mg with little change in soil pH. Similarly, if additional Ca is required but the pH is already high, then calcium sulfate can provide additional Ca. Evidence also exists that Ca ions can suppress *Phytophthora* disease of raspberries. Pre-plant applications of calcium sulfate have been shown to reduce root rotting in *Phytophthora*-infested fields of red raspberry (Maloney *et al.*, 2005).

Soil pH may also be affected by the source of N, but this comes into play only after the plants are established. Nitrogen fertilizers should not be used to modify pH before planting. Other nutrient sources have only a small impact on soil pH. Therefore, lime and sulfur are the basic tools for managing soil pH.

A standard soil test analysis also reports the estimated amount of plant available P and K, which will vary depending on the extractant used. If additions are recommended, then follow the recommendation of a lab or consultant who has experience with raspberries and blackberries. As is the case with lime and S, the more finely ground the source of P and K, the faster it will become available to plants. In addition, sources that are readily soluble in water will be available faster than those that are less soluble. Organic sources of P and K, whether occurring naturally in the soil or provided through organic fertilizers, tend to have nutrients tied up in complex mineral structures that are only slowly released to the plant. Conventional inorganic fertilizers provide a more readily available form of nutrient, but they contribute saltiness to the soil (e.g. chlorides, sulfates) that can negatively affect plant growth if their levels are too high. A combination of organic and conventional sources is often desirable.

Many soils are sufficiently high in P for raspberries, especially if they have received manure applications in the past or have had annual field crops grown on them. Excessively high P can interfere with uptake of certain other essential nutrients, so P should not be applied without first testing the soil to determine if it is needed.

In areas that typically have low B, it is a good idea to test for it, even though this usually incurs a separate charge from the laboratory. Boron affects root development, so low B can reduce uptake of all other nutrients even if there are adequate levels of these nutrients in the soil. Boron also has profound effects on bud break, and in deficient soils, bud break and production may be severely reduced.

Soil test values for other nutrients, especially micronutrients and N, do not provide a reliable basis for making recommendations. Adjustments to these nutrients are best made once plants are established and leaves can be used as an indicator of nutritional status.

All amendments should be worked into the root zone prior to planting. Applying nutrients to the soil surface after plants are established is not very effective because most fertilizers move slowly through the soil profile. Incorporating lime, S and nutrients several months before planting will allow sufficient time for appropriate changes to take place in soil chemistry.

N is an essential nutrient for plant growth and is generally considered desirable in soils. Higher soil N usually means that the soil is fertile. It is available in the soil in three basic forms: ammonium N, nitrate N and organic N. Raspberries preferentially use nitrate N. Typically, nitrate forms of N are low in soil, so they are usually added through fertilizers. Newly planted raspberries do not require large amounts of nitrate N, so it is best if it can be supplied by converting from the other sources already present in the soil. Ammonium N is converted to available nitrate N when soils are warm and the pH is not too low. Organic N is converted to available nitrate N when biological activity is high and there is not too much carbon in the soil. The best method of providing a pool of nitrate N to plants prior to planting is by incorporating a source of organic N that is low in carbon. Examples include manure or a legume cover crop. These materials break down slowly and provide a source of nitrate N to the plants without significant leaching. In situations where organic N is low, supplemental nitrate fertilizer may need to be provided, but this rapidly available form is subject to leaching. Too much nitrate or ammonium N fertilizer can be toxic to plants. It would be rare for organic matter to provide excessive N to the plant, except in situations where it is released late in the season as plants attempt to harden off for winter. In most situations, increasing organic matter prior to planting is a valuable step in the soil management process.

Other chemical elements can be toxic to plants and people, so soils high in these elements should be avoided. Arsenic (As), Se, As, Na and Al are naturally part of certain soils, but when they are taken up by plants in large amounts, detrimental responses result. Uptake of these elements is usually increased as acidity increases, so pH should be maintained above 6.0 if these elements are present at high levels. Excessive amounts of essential nutrients can be toxic as well, so it is critical not to over-fertilize. Certain fertilizers may contain other elements that are toxic in high amounts. For example, muriate of potash (potassium chloride) is a useful fertilizer for raspberries so long as amounts are low. If applied amounts are high, the excessive chloride in the fertilizer can be harmful if there is not sufficient rainfall to leach the chlorides. In such cases, potassium sulfate is recommended even though it is more expensive.

A second potential source of toxic chemicals in soils is irrigation water. Often the toxicity is caused by too much salt in the water. This is especially problematic where drip irrigation is used under dry conditions. Test the irrigation water for salt content before using. It should be less than 2.0 dS/m, preferably less than 1.0.

A third potential source of toxicity is herbicides that have been used for weed control in previously planted crops. Laboratory testing for herbicide residue is inexact and expensive, so a bioassay is recommended. A bioassay is simply a comparison of the growth of several types of seeds (e.g. radish, bean, rye, cucumber) planted into a sample of soil from the intended field and in a similar soil that has never been treated with herbicide. If growth differences occur or injury symptoms are expressed, then it might be attributed to herbicide residue. If a difference exists, waiting another year before planting may be prudent.

SOIL QUALITY EVALUATION

Recently soil scientists have been developing evaluations of soil health that consist of a representative set of chemical, physical and biological variables. Traditionally, only chemical assessments have been made because they are easy to obtain. However, physical and biological properties of soils also play a large part in plant performance. Chemical assessments are provided by standard soil tests. Among the many physical properties of soils, the following are good indicators of soil health: aggregate stability, available water-holding capacity, surface and subsurface hardness and soil texture. Good biological indicators are soil organic matter content, active carbon content (the quality and type of organic matter), potentially mineralizable N (ability of organic matter to supply N) and root health (presence of plant disease and nematodes). One assumes that there is cause-and-effect between these soil quality variables and plant performance; therefore, improving any of these should eventually result in improved plant performance.

SITE PREPARATION

Preparing a site for planting requires a thoughtful plan to create an environment in which plants perform to their maximum potential. Select a site with an appropriate soil type – in particular, the soil should have adequate internal drainage. If the site has standing water, it should be drained. If drainage is not possible, then plan on planting on raised beds of at least 25 cm (10 in) in height. Perennial weeds should be suppressed or killed at least a year before the intended planting date. This may require two or more applications of a broad-spectrum, post-emergent, non-selective, non-residual herbicide.

Conduct soil and nematode tests the summer before planting and incorporate recommended soil amendments to a depth of 20 cm (8 in) if possible. If the location was treated with a residual herbicide recently, then conduct a bioassay to ensure that residues will not impact crop growth. Grow a sequence of cover crops one year or more before planting, including some legumes, and use minimal tillage to incorporate them.

Preparing a site for raspberries does not require special equipment. Applying a broad-spectrum herbicide requires a boom sprayer. Plowing, disking, applying lime or fertilizer and tilling require a tractor with a plow, disc, spreader and tiller attachment. These are all commonly used for site preparation for most crops.

By following these steps, the site should be ready for planting the following spring. The soil pH and nutrient levels should be adequate for the life of the planting. Drainage, if necessary, has been installed. Perennial weeds should have been eliminated and nematodes suppressed. Organic matter levels should have increased, along with biological activity and water-holding capacity. Soil tilth and aggregation are preserved as much as possible. Irrigation water has been tested and is low in salts with a neutral or slightly acid pH.

FIELD LAYOUT

On flat ground it is generally better to plant rows oriented north–south rather than east–west as light interception by plant rows will be greater. Because the sun rises in the east and sets in the west, the entire canopy receives exposure to direct sunlight at some point during the day (as long as rows are not too close together). The only exception is if fall raspberries are being grown in a high tunnel for late season production. In this case, east–west rows intercept more sunlight when the sun is low in the western sky after the equinox.

The aspect of the slope is also an important consideration when the goal is to maximize light interception of the canopy. Steep, north-facing slopes at high latitudes may never receive direct sunlight and plants may not grow well under these conditions. In contrast, in the northern hemisphere during summer, plants on a south-facing slope will receive more direct sunlight than those facing other directions. Differences in light interception are small when the angle of the slope is small, but become significant when the angle of the slope increases, especially at higher latitudes.

Raspberries are particularly sensitive to wind damage. Even though damage may not result in visible symptoms, raspberries grown under windy conditions generally have shorter canes and lower yields. Two strategies are used to mitigate wind damage. The first is to use natural or artificial shelter belts to break the wind. Typically, if wind comes from the west, a north–south windbreak on the west end of the field will help reduce wind damage. The windier the site, the more frequent these windbreaks should occur. Raspberry rows themselves may be less susceptible to wind damage if planted in the direction of the prevailing wind, but light interception must be considered as well. The taller the windbreak, the longer the distance that wind speed will be reduced.

A second strategy to mitigate wind damage is to use a trellis. All brambles benefit from a trellis, even those grown for a fall-crop only. A trellis not only helps prevent canes from moving in wind and damaging their vascular connections with the root, but also holds canes erect, helps improve exposure of fruit, and improves efficiency of harvest and fruit quality. Trellises for fall-fruiting types need not be elaborate, but simply need to hold canes erect. Trellises for floricane-fruiting plants need to be more rugged to hold primocanes and fruit-bearing canes erect throughout the year.

Materials for constructing trellises have traditionally consisted of 5 × 10 cm (2 × 4 in), or 10 × 10 cm (4 × 4 in), treated posts, with metal wires strung between posts. Posts are usually 8–10 m (25–30 ft) apart down the row, sometimes with metal stakes in between. Floricane-fruiting types benefit from having the canes spread into a V-shape so that light interception and penetration, especially into the lower canopy, is improved. Newer, lighter-weight materials have been developed for trellises. Monofilament plastic wire is as strong as steel wire, and is easier to work with as it does not conduct electricity in the event of a lightning strike. Fiberglass posts are easy to adjust and just as strong as steel.

PLANTING

Several choices of planting stock are available, depending on the nursery. The traditional and most common type of red raspberry are the bare root suckers taken from mother plants in a nursery. These have a short piece of cane attached to a rather large root and are dug dormant in fall. Bare-rooted plants are kept in a cooler until shipping in spring. They can be planted in spring as soon as the soil can be worked. Black raspberries can be propagated by root tips. They, too, are dug in the fall with roots attached to a shoot, so are best transplanted in early spring.

Tissue-cultured green plants are similar in size to a pepper or tomato transplant, and so are much smaller than bare-rooted plants. The plugs are easily handled by most mechanical transplanters. They should be planted only after the danger of heavy frost is over because the leaves are green and sensitive to extreme cold.

Green tissue-cultured plugs can be chilled in the nursery or subjected to conditions that encourage leaf drop and dormancy. These propagules are very small – just a root ball with a leafless shoot, a few centimeters tall. They can be planted while there is still a risk of frost. These dormant tissue-cultured plugs ship well because they have no leaves.

Studies have compared the performance of these three types of propagules. By the time these plants reach maturity, the performance is about equal among the propagule types. Therefore, the choice of which to purchase is based primarily on characteristics other than yield. Green plugs might be the best choice if the planting date is after frost and irrigation water is readily available. Dormant, bare-root plants might be the best choice if no irrigation is available.

Typically, raspberries are set in rows, with the distance between rows depending on the plant type and their tendency to produce laterals. Between-row spacing is a minimum of 3 m (10 ft), while the intra-row spacing is from 30 cm to 1 m (1–3 ft), depending on the training system. Primocane-fruiting raspberries can be planted as close as 2.6 m (9 ft) between rows and 30 cm (1 ft) between plants within rows. In high tunnels and greenhouses where tractors are not used down rows, between-row spacing can be as close as 1.8–2 m (6–7 ft).

Studies have shown that shallow-rooted plugs are sensitive to herbicides and soil disturbance after planting (Neal *et al.*, 1990). Therefore, planting through plastic mulch or mulching with straw or newspaper after planting helps suppress weeds while retaining soil moisture (Warmund *et al.*, 1995; Percival *et al.*, 1998). Plants mulched their first year tend to outperform unmulched plants because herbicide use and cultivation are avoided and moisture levels remain consistently high (Trinka and Pritts, 1992). Plants can also be set into killed rye or other cover crop, mimicking a no-till or strip-till situation. This strategy helps suppress weeds while the plants become established without a lot of soil disturbance.

Once plants are set, they should be irrigated. The shallow root system of plug plants cannot access much soil moisture; therefore, sufficient water should be applied to maintain some moisture in the root zone of the plants.

Once plants are set, mulched and irrigated, they are well on their way to becoming established. If recommended procedures are followed, fertilizer should not be required because nutrients should have been incorporated into the soil during the pre-plant step, and sufficient N will be provided from organic matter for several months after planting. The most important function for the grower after planting is to keep weeds under control. If the planting establishes well and weeds are kept under control the first year, then it will be difficult for weeds to invade the planting later.

In many cases, a perennial grass is planted between rows of berries to help suppress weeds and provide a suitable surface for tolerating equipment and movement of foot traffic, especially after it rains. Fescues and dwarf perennial ryes are good choices. Grass seed germination is best when soil is moist and temperatures are cool, so fall planting is often the best time to seed row middles.

An initial investment in the pre-plant site preparation and modification can pay major dividends for growers in the ensuing years, but planning ahead is the key to success.

REFERENCES

Belair, G. (1991) Effects of preplant soil fumigation on nematode population densities and on growth and yield of raspberry. *Phytoprotection* 72, 21–25.

Black, B.L., Swartz, H.J., Millner, P. and Steiner, P. (2003) Pre-plant crop rotation and compost amendments for improving establishment of red raspberry. *Journal of the American Pomological Society* 57, 149–156.

Forge, T.A., Walters, T.W., Koch, C.A. and Particka, M. (2009) Effects of compost and manure mulches on soil biological activity, population densities of *Pratylenchus penetrans*, root-associated fungi, and root biomass of 'Meeker' red raspberry. *Canadian Journal of Plant Pathology* 31, 136.

Hargreaves, J., Adl, M.S., Warman, P.R. and Vasantha Rupasinghe, H.P. (2008) The effects of organic amendments on mineral element uptake and fruit quality of raspberries. *Plant Soil* 308, 213–226.

Maloney, K., Pritts, M., Wilcox, W. and Kelly, M. (2005) Suppression of Phytophthora root rot in red raspberries with cultural practices and soil amendments. *HortScience* 40, 1790–1795.

Neal, J.C., Pritts, M.P. and Senesac, A.F. (1990) Evaluations of preemergent herbicide phytotoxicity to tissue culture propagated 'Heritage' red raspberry. *Journal of the American Society for Horticultural Science* 115, 416–422.

Orhan, E., Esitken, A., Ercisli, S., Turan, M. and Sahin, F. (2006) Effects of plant growth promoting rhizobacteria (PGPR) on yield, growth and nutrient contents in organically growing raspberry. *Scientia Horticulturae* 111, 38–43.

Percival, D.C., Proctor, J.T.A. and Sullivan, J.A. (1998) Supplementary irrigation and mulch benefit the establishment of 'Heritage' primocane-fruiting raspberry. *Journal of the American Society for Horticultural Science* 123, 518–523.

Pinkerton, J.N., Ivors, K.L., Reeser, P.W., Bristow, P.R. and Windom, G.E. (2002) The use of soil solarization for the management of soilborne plant pathogens in strawberry and red raspberry production. *Plant Disease* 86, 645–651.

Seigies, A.T., Pritts, M.P. and Kelly, M.J. (2006) Cover crop rotations alter soil microbiology and reduce replant disorders in strawberry. *HortScience* 41, 1303–1308.

Taylor, J. and Harrier, L. (2000) A comparison of nine species of arbuscular mycorrhizal fungi on the development and nutrition of micropropagated *Rubus idaeus* L. cv. Glen Prosen (Red Raspberry). *Plant Soil* 225, 53–61.

Toussaint, V., Valois, D., Dodier, M., Faucher, E., Dery, C., Brzezinski, R., Ruest, L. and Beaulieu, C. (1997) Characterization of actinomycetes antagonistic to *Phytophthora fragariae* var. rubi, the causal agent of raspberry root rot. *Phytoprotection* 78, 43–51.

Trinka, D.L. and Pritts, M.P. (1992) Micropropagated raspberry plant establishment as influenced by weed control practice, row cover use and fertilizer placement. *Journal of the American Society for Horticultural Science* 117, 874–880.

Warmund, M.R., Starbuck, C.J. and Finn, C.E. (1995) Micropropagated 'Redwing' raspberry plants mulched with recycled newspaper produce greater yields than those grown with black polyethylene. *Journal of Small Fruit and Viticulture* 3, 63–73.

Wilcox, W.F., Pritts, M.P. and Kelly, M.J. (1999) Integrated control of Phytophthora root rot of red raspberry. *Plant Disease* 83, 1149–1154.

SOIL AND WATER MANAGEMENT

RICHARD C. FUNT[1]* AND DAVID S. ROSS[2]

[1]*Department of Horticulture and Crop Science, The Ohio State University, USA;* [2]*Department of Environmental Science and Technology, University of Maryland, Maryland, USA*

INTRODUCTION

Each grower must intelligently weigh certain fundamentals in site and soil selection. Success in the raspberry business is dependent on the best management of the land and water resources. The farm may grow different crops for different markets, and therefore, requires a management plan for the location of the infrastructure for the water supply and how water is to be applied to different soil types and different crops. A site should be chosen with the soil characteristics needed for the best long-term raspberry production (Slate *et al.*, 1949).

An effective management plan should include the future expansion of the business in regards to water usage, movement of vehicles across main irrigation lines, size of irrigation equipment and the rotation and/or expansion of crops. These need to be oriented to the type of production and marketing system that the grower selects. In general, the grower needs to have both a long-term and a short-term economic planning horizon to be effective and efficient with an operation. Soil and water management over the short and long term is very important.

The primary goals of soil management are to maintain fertility, reduce erosion of the topsoil and maintain good soil structure with minimum tillage. The primary goal of water management is to maintain adequate soil moisture for producing an optimal canopy for maximum fruit production. Water also needs to be supplied from bloom to harvest for maximum yield and fruit size, but without leaching nutrients into the water table. Also, the management of soil fertility and soil moisture will produce large canes for the current or subsequent year's crop.

*richardfunt@sbcglobal.net

SOIL MANAGEMENT, FERTILITY AND FERTILIZATION

Soil management systems used in raspberry production can be defined as clean culture, herbicide strips with grass (sod) culture, herbicide strip with cover crop system, and a combination of grass culture and woven weed barrier. The clean culture system refers to hand weeding or a combination of cultivation using mechanical hoes, disks or harrows, and hand weeding. In Maryland between 1890 and 1930, black raspberry growers used hand weeding and cultivation using horse-drawn cultivators (Ross and Auchter, 1930). The fields were weeded three to four times per year. To minimize damage to the raspberry roots, cultivation was shallow and/or kept some distance from the crown. Animal manures were used, increasing the quantity of weeds, as seeds in the straw or hay bedding used for the animals got into the field. Cultivation, during and soon after harvest, loosened the soil which was packed down by the people (pickers) who harvested the fruit. Further, cultivation during harvest could result in dropped fruit. Economically, this system over time destroys soil structure and requires many hours of labor.

During the 1970s, growers in the eastern USA converted to using herbicides in the row and using grass (sod) as a permanent ground cover in the row middle or drive row. This reduced hand weeding labor by 80% and reduced the destruction of soil structure, improved organic matter, and reduced erosion. Certain herbicides are used immediately after planting so as to not harm young roots. Different herbicides are recommended for year three and older plantings.

Raspberries require open, porous and slightly acidic soil (pH of 6.0–6.5) for good growth, movement of water and good nutrition availability. A balance of calcium, potassium and magnesium is necessary for optimal plant growth, optimal yields and the best fruit quality. Water from rainfall and supplemental irrigation is necessary to allow the transfer of nutrients from the soil to the plant roots. The ideal soil is well drained (good percolation below the root zone); a sandy or silt loam soil with good moisture retention and an organic matter content of 2–4%. Poor drainage can result in problems with root rots, the cause of much winter injury. The wet soil stimulates late autumn growth and the plant fails to harden off properly. Raspberry roots are generally in the top 60 cm (2 ft) of the soil. Thus, the water level should not come within 1 m (3 ft) of the soil surface for more than a few days. Where soil water percolation may be marginal, soils can be modified slightly to correct some problems. Raised beds that are 20–25 cm high (8–10 in) can reduce root rot problems (Funt and Bierman, 2000). Raised beds are drier than flat surfaces and will require more frequent irrigation (Fig. 9.1).

Soils can be modified with organic matter (humus), such as green manure, animal manure and plant or animal based composts. This will improve the moisture-holding capacity, the fertility and the cation exchange

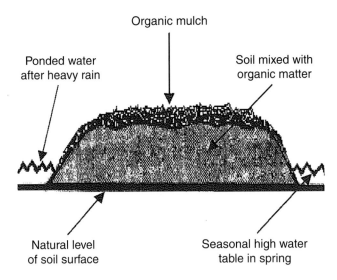

Fig. 9.1. Raised bed for water management of raspberries.

capacity (CEC) of the soil. A silty clay loam soil or a sandy clay soil, well supplied with organic matter, is desired and hardens less than those with a lower level of organic matter. Adding organic matter, such as leaf compost, to a raised bed and incorporating it into the top 10 cm (4 in) creates an improved rooting environment for plant growth.

Soils should be tested before planting. Any major or minor nutrients/minerals that are deficient should be incorporated into the upper 30 cm (12 in) of soil before planting. Soil fertility, gas exchange (oxygen), pH and CEC may be improved with organic matter. Organic matter can be incorporated into the soil prior to planting (refer to Chapter 8).

The CEC is generally related to the soil texture; clay soils have a higher CEC than sandy soils because they have a higher total particle surface area than sandy soils. One to 2 years prior to planting, raspberry growers should take a soil sample to determine the amount of sand, silt and clay (soil texture), organic matter and CEC. Materials should be applied to improve the soil if required (refer to Chapter 8).

After the soil has been prepared and the plants have been planted, the amount of fertilizer for optimal production and winter hardiness is a major concern. In general, phosphorus, potassium, calcium and magnesium should not be required for the first several years. However, a soil test 1–2 years after planting may be necessary to maintain sufficient levels. Raspberries can be heavy feeders of nutrients, such as nitrogen (N), potassium (K), and possibly magnesium (Mg) and phosphorus (P), depending on soil type and climate. Maintaining a pH of 6.0–6.5 can aid in uptake of nutrients and reduce the likelihood of deficiencies. Therefore, the importance of applying the necessary nutrients before planting and incorporating them into the soil cannot be overemphasized. After planting, only N may be needed for the first several years. Additional amounts of K may be necessary in years 3 or 4. Foliar sprays of minor elements, such as boron (B), can be helpful under wet (slow percolation) soil conditions in the spring after bud break when the leaves are beginning to show.

Soil applied N, P, Mg and calcium (Ca) are cheaper than, and just as effective as, foliar applications. However, some research indicates that new and more effective foliar nutrients may be beneficial when applied in early autumn just before leaf fall and before a frost. Fertilizers injected into drip irrigation systems (fertigation) do show promise in being efficient, plus the amount used is much less than with materials applied to the soil (Fig. 9.2). Caution is advised in that certain fertilizers can clog filters and emitters in drip systems. In certain areas in the USA growers are required by the government to create a nutrient management plan and to report what nutrients were applied and in what amounts.

Certain grasses planted in the sod drive row can reduce soil compaction when farm equipment is used under wet soil conditions. In areas of slow soil

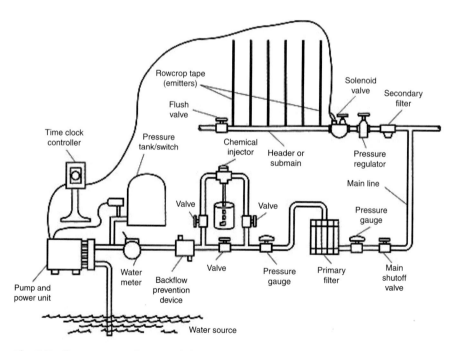

Fig. 9.2. Components of a typical drip irrigation system for raspberries. (Source: Ross, 2004.)

water percolation with clay-type soils and during periods of heavy rainfall, grasses such as tall fescue can aid equipment travel between the rows for applying pesticides and/or the use of mechanical harvesters. Also, sod is mowed and creates a solid pathway for pickers. Overall, this reduces soil compaction, allows water infiltration under rapid rainfall, reduces erosion on hillsides and reduces weeds (biological weed control) between rows.

New cultivars of tall fescues include dwarf types, which are drought tolerant, require fewer mowings (less trash for mice and voles) and are usually 'endophytic', unless the label specifies they are endophyte-free. Endophytes are also found in certain perennial rye grasses, tall chewing or hard fescues, and improve sod survival. Non-endophytic types do not do well under difficult conditions when the seeds become infected with a fungus (*Neotyphodium coenophialum*), which creates a toxin for insects and nematodes. Also, the endophyte infection causes early closing of the stomata to conserve moisture during droughts. Early closing of the stomata also conserves nutrients. These chemicals can reduce chinch bugs, thrips, bill bugs, sod webworms and aphids that feed on grass stems. The endophyte also aids the fescues in storing more organic carbon (global warming) and nitrogen in the soil. Because endophytes also produce substances toxic to livestock, cattle or other animals should not be allowed to eat these types of fescues or grasses.

Another system in use is an inter-row cover crop that is killed by winter conditions, by cultivation or mowing, or by the use of herbicides. A seasonal cover crop, such as oats, is planted between the rows, usually in the summer, and is incorporated into the soil the following spring. This improves organic matter, improves soil structure and increases soil organic matter; it can also be used to reduce levels of nematodes and other pests. A herbicide is used within the row of raspberries. Seasonal cover crops can also reduce excess nitrogen. In very fertile soils in the north-eastern and Midwestern USA, raspberries grow too vigorously and too late in the fall, which can result in early freeze injury. Further, when the cover crop is incorporated into the soil in the spring, the nitrogen will be released back into the soil, reducing the need for nitrogen fertilizer.

Raspberries grown in a high tunnel can be grown in any of the systems mentioned above, although in four-season tunnels, sod cannot be grown between the rows without irrigation. In greenhouses (glasshouses), soil or potting media can include other non-soil derived materials, such as sand, peat moss or vermiculite mixtures plus nutrients for optimal plant growth, either for in-ground or pot culture.

Raspberry growers may want to consider specialty buildings or shelters, such as a greenhouse or high tunnels. Greenhouses are considered to be permanent structures and use heat and electricity. They may be used for berry production. In many operations, these structures are used for annual bedding plant production or vegetable starter plants. Raspberry producers should consider a design for the placement of structures near water supplies, electrical sources and waste management. New systems can recycle water or water-containing fertilizers, reducing water and fertilizer usage.

In Ohio, where 60% of the soil requires improvements in internal drainage, raised beds have been recommended for raspberries. Generally, the top layer of soil (10–15 cm/4–6 in) is a silt loam but going deeper it can be a clay loam, which is slow in water percolation. By placing an additional layer (10–15 cm/4–6 in) of topsoil in the planting row, with a one- or two-bottom plow, the grower can create an environment of 12–25 cm (5–10 in) of well-drained soil. Loosening the soil, with deep rototilling prior to making the beds in dry soil, can provide greater success in heavy soils. Generally, beds are made in autumn and prepared for planting in spring, although in milder climates autumn planting with dormant plants may be favored for facilitating herbicide control of weeds. A soil test should be completed and, based on the soil test results, nutrients such as Ca, K, zinc (Zn) and P should then be incorporated into the top 10–13 cm (4–5 in) with a rototiller. However, deep rototilling is not advised because it will destroy or flatten the raised bed.

Many raspberry plantings are grown in soils where peach and cherry trees can be grown. These soils tend to be fertile, well drained and have porous structures with ample oxygen for roots in the upper 20–30 cm (8–12 in). Further, the Puyallup Valley in Washington State probably represents an intensive small fruit district in the USA and provides good soil tilth. In Washington County, Maryland, black raspberries are grown on gravelly loams; the rocks in these soils include sand, slaty and shaly types. The heavier soils in this area and in Oregon are well drained and seem well adapted to berry growing.

Root health is increased with improved soil structure from the application of organic matter; and therefore, increased pore space, thus increasing oxygen and gas exchange and internal water drainage. In Ohio, on Crosby silt loam, composted yard waste decreased soil bulk density, decreased water-filled pore space, and therefore increased porosity of the soil by 10–40%. When compost was incorporated into the soil and applied to the surface of the soil, soil water holding capacity and the rate of infiltration was increased by 36–40% and was 7 to 21 times faster, respectively, in the first 2.5 cm (1 in) depth as compared to the non-treated soils (Funt and Bierman, 2000). Raised beds allow better soil water drainage under heavy prolonged rainfall. However, raised beds also dry out much faster and to a greater depth than flat (non-raised) beds (Fig. 9.1). Thus, irrigation becomes a part of the raised bed system of culture. Further, when the highly susceptible root rot cultivars, such as 'Titan', were planted on a silt loam soil containing *Phytophthora fragariae*, raised beds (35 cm/14 in high) dramatically reduced the incidence and severity relative to flat beds (Maloney *et al.*, 1993). Accordingly, the reduction in disease severity provided by the raised bed was due to the provision of a rooting zone in which the soil water tension generally exceeded that which supported zoospore activity. With all these factors, plants are healthier and can resist more insect and disease pressure than plants grown with lower amounts of nutrients, organic matter or porosity soil.

IRRIGATION AND WATER MANAGEMENT

The quantity of water in raspberry production is important throughout the life of the planting. Too much water over a long period of time on poorly drained soil will cause plant root disease and plant loss. Improving internal soil water drainage and porosity will greatly enhance the life of the raspberry planting. Because the raspberry roots (see Chapter 2) are relatively shallow, too little water causes plant water stress and poor plant growth, regardless of the soil fertility and internal drainage. Reduced amounts of water over the short and long term decrease shoot growth for the current season and reduce berry size and decrease fruit set for the next season's crop.

Water supply

The irrigation water supply must be large enough to meet the needs of the total enterprise. Large ponds, lakes, streams, springs, groundwater, municipal water and waste water are all potential water sources. These may be supplemented by manmade reservoirs built to collect run-off or to store water received steadily over time to allow an irrigation application (Ross and Wolf, 2008). In some areas, permits are required before surface water or high-capacity wells can be developed and/or used for irrigation. In general, when rainfall does not occur for several weeks the planting is under drought conditions. Water sources need to hold or be able to supply sufficient water for all crops and must be functional in droughty seasons. A storage reservoir can be charged (filled) from an alternate source, such as a well or stream.

A 2.5 cm (1 in) depth of water over the surface of 0.4 ha (1 acre) contains 102,780 l (27,154 gal) of water. This is equivalent to about 256,940 l/ha for a 2.5 cm depth. Under average conditions, this amount of water may be required by the crop each week. The water supply must be able to satisfy the most demanding criteria. Overhead, flood or drip irrigation systems are widely used in various parts of the world. Further, climatic conditions (normal weather, heavy rains or drought) determine water needs.

When calculating the amount of water needed for raspberries using drip irrigation, only the root zone area needs to be irrigated. Drip irrigation applies water to a part of the root zone and does not broadcast the water as an overhead or flood method does, so it allows more efficient application of water to the desired crop. Assuming a crop root zone width of 0.9 m (3 ft) times 90 m (300 ft) length of row, the equivalent area to wet is 0.008 ha

(900 ft² or 0.02 acre). To apply 2.5 cm (1 in) of water requires 2060 l of water (543 gal). One must estimate how many times in a season this amount of water will be applied in order to estimate the seasonal requirement for the crop. Typically, about 0.25–0.75 cm (0.1–0.3 in) per day of water in the root zone is required by the crop over the season. Some allowance must be added to account for the efficiency of the application of irrigation water due to evaporation.

The total volume of required water might be estimated over a growing season based on an estimated water use curve (average daily water use in inches per day × number of days) and area to be irrigated. Generally, it is most critical to understand what the water supply demand will be greatest during the peak demand periods of the growing season and to understand the effects of a long drought on the water supply. Some streams and rivers can be very low during dry periods. Ground water table levels can drop during a drought. Learn about the potential water supplies of your growing area before making plans for crop production.

In many farming areas, the amount of water available for irrigation is limited to less than is required to operate an overhead or furrow system. Drip irrigation becomes the only option. Drip irrigation can work with water supplies as low as 185 l (50 gal) per hour per acre or less than 3.8 l (1 gal) per minute, but that is not much water. In some cases water must be pumped or received 24 hours a day into storage tanks or holding ponds to irrigate a small area during the day. The water supply must be evaluated to determine whether it can meet irrigation requirements.

Water quality

Good water quality is defined as having a suitable pH, from 5.5 to 7.0; being free from dissolved salts; being free from soil particles, such as sand, silt or clay, and being free of biological materials, such as plants (algae) and waste products containing *E. coli*, heavy metals (cadmium, lead) and pesticides (Ross, 2004). Good quality water is required for drip irrigation systems. Generally, the highest water quality comes from wells and/or springs that are protected against contaminants. Well water is used for drip irrigation systems in fields, greenhouses and high tunnels. Water may be obtained from streams (when authorized), irrigation canals or ditches, or reservoirs, ponds or lakes and applied with overhead or furrow types of irrigation, including permanent pipes or center pivot or traveling guns. These sources offer larger amounts of water per minute but may contain soil, algae and other materials. Filters will be required to remove these materials so that drip emitters or sprinkler nozzles are not clogged. After application, monitoring of soil water quantity and quality is also beneficial in order to schedule irrigations. A commercial water test should be obtained initially and then every year or two to establish the stability of the water quality. Testing avoids finding trouble later that might require costly treatment.

Water requirements

The amount of water held by the soil (water-holding capacity) is a factor of the amount of clay and organic matter. An understanding of the hydrological cycle is necessary for a grower to make soil water management decisions (Fig. 9.3). When rainfall or irrigation exceeds more than the infiltration and

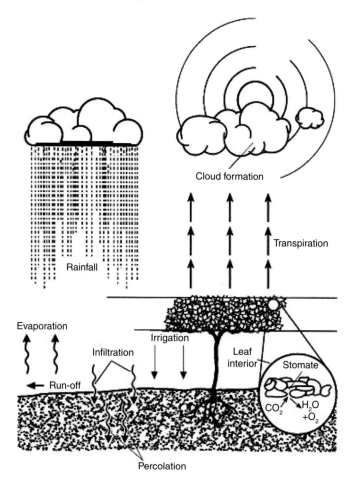

Fig. 9.3. The raspberry hydrological cycle. Water enters the vineyard as rainfall or irrigation and is removed thorough gravity, run-off, evaporation and transpiration through plant leaves. (Courtesy of Wine Grape Production Guide for Eastern North America. NRAES 145.)

percolation rates of a soil (i.e. 2.5 cm (1 in) per hour for well-drained soil), water runs off the surface and can cause erosion and loss of nutrients (Ross and Wolf, 2008). Grasses between rows can reduce erosion, as explained above. Contour planting can help reduce run-off and erosion.

Twenty-eight liters (1 ft^3) of nearly ideal soil contains 50% soil particles, 25% moisture (about 7.5 cm (3 in) or about 7 l (2 gal) of water) and 25% air space, if one envisioned these as parts of a cube. Different soil types hold different amounts of water (Table 9.1). The rooting depth of the raspberry can vary depending on the soil type and internal soil drainage, but the fibrous roots which take up moisture and nutrients are mainly in the upper 30 cm (1 ft) of soil. At 0.5 cm (0.2 in) of water loss per day, the plant could be near water stress when more than 50% of the available soil water has been used by the plant or evaporated in about 4–5 days. This amount is also referred to as the plant available water (PAW) and is the amount of water between field capacity and the permanent wilting point. Field capacity is the amount of water the soil can hold after it is saturated and allowed to drain away the water not held by soil particles. The permanent wilting point is the amount of water held so tightly by the soil that plants cannot extract it. The goal is to maintain the PAW at 50% or higher so plants are not stressed for lack of water.

Further, the presence of a grass (sod) alley and/or raised bed will affect the amount of water needed to produce a crop of raspberries if overhead irrigation is used. In warm climates, the amount of water needed could range from 2.5 to 3 cm (1.0–1.3 in) per week for the crop and when the additional 30–40% air evaporative loss is calculated during mid-summer, the total water loss could be 3–5 cm (1.3–2.0 in) of water per week. Drip irrigation, however, puts the water in the root zone of the raspberries so there is much less influence by the grass alley or air evaporation and only the crop requirements must be met.

Table 9.1. Water-holding capacity of different soil textures. (Source: Bramble Production Guide, NRAES-35.)

Soil texture	Available water-holding capacity, cm of water/cm of soil (in of water/in of soil)
Course sand	0.02–0.06
Fine sand	0.04–0.09
Loamy sand	0.06–0.12
Sandy loam	0.11–0.15
Fine sandy loam	0.14–0.18
Loam and silt loam	0.17–0.23
Clay loam and silty clay loam	0.14–0.21
Silty clay and clay	0.13–0.18

During the spring there could be an excess amount of water received, making raised beds beneficial in an area with poorly drained soils. Raised beds during the harvest period (Fig. 9.1) are much drier than non-raised beds. Raspberry plants on raised beds suffer less winter injury than plants on flat areas because there is less chance of roots being in saturated soil.

A general recommendation is to design an irrigation system to replace the water lost during the months of greatest water loss (July and August in eastern USA) plus 25% for extreme hot and windy conditions (Fig. 9.4). The daily loss of an average day in the humid eastern USA ranges from 0.45 cm to 0.62 cm (0.2–0.25 in) per day. In California and Washington State that rate is as much as 0.8–0.95 cm (0.3–0.35 in) per day. A general recommendation for drip irrigation of raspberries is to apply the amount of water to the root zone of the plant according to the evaporative loss plus whatever weather conditions occurred over the past 24 h. Water should be applied, unless there has been 2.5 cm (1 in) of rainfall during that period of time. Applying water in the morning can be effective in reducing water or leaf stress and helps to cool the plant (Fig. 9.5). Raspberry growers should invest in soil moisture monitoring equipment to understand the quantity of water needed to reach peak plant performance.

Moisture monitoring

Irrigation management is best done by monitoring the soil moisture in the root zone under the irrigation system on a frequent basis and providing water to maintain a reasonable plant-available amount of water. Trying to estimate water usage by daily evaporation bookkeeping methods or manually reading tensiometers and other sensors is being rapidly replaced

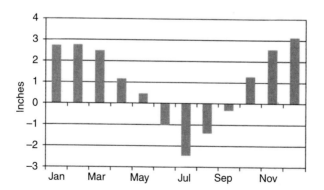

Fig. 9.4. The imbalance between precipitation and potential evapotranspiration is illustrated for Washington Dulles International airport in northern Virginia (1 in of rainfall = 25 mm of rainfall). (Source: http://climate.virginia.edu.)

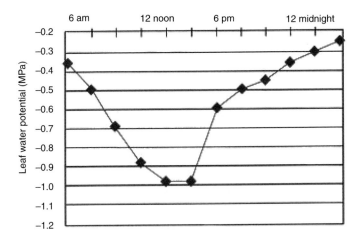

Fig. 9.5. Changes in leaf water potential throughout the course of a day. (Degree of leaf stress is indicated by increasingly negative values.)

by newer technologies. Newer capacitance sensors are available that can be calibrated to specific soils or growing substrates to give accurate volumetric moisture content and relationships to moisture tension, the measure of water availability to plants (Lea-Cox et al., 2008). Real-time data can now be obtained for scheduling irrigations to maintain uniform soil moisture conditions for plant growth.

In late 2010 technology, moisture sensors attached to wireless radio nodes (battery powered) send the moisture data (volumetric moisture content) on a frequent basis to a wireless radio node attached to a computer in the office. Software in the computer stores the data and graphically displays it for the grower to make irrigation scheduling decisions in real time. The wireless radio nodes in the field create a network to relay the data via line of sight so node-to-node transmission pathways can change if something blocks the pathway. Battery life of the radio nodes lasts for the full growing season. Also, some current moisture sensors can also measure soil/substrate temperature and electrical conductivity. Weather station sensors (i.e. rainfall, air temperature, photosynthetically active radiation (PAR) light, leaf moisture, and air velocity and direction) are also connected to wireless radio nodes to send real-time data to be graphically displayed. In development and testing are wireless switching radio nodes with microprocessors that can collect local data, evaluate it, and based on set points, turn on and off irrigation solenoid valves in an irrigation zone (D.S. Ross, 2010, personal communication).

This technology will give the grower much more information about their growing conditions to allow for better management. It is true that one must learn where to place the sensors to get the most meaningful data but this can be quickly established. The systems are portable for trial and error movement. Growers who have experience with the sensor networks consider them essential for efficient moisture management and for understanding what is happening in the crop environment.

Make an irrigation business plan

Raspberry growers need to have a business/management plan for designing the irrigation system for the present and future needs (Fig. 9.6). An overall plan and design can reduce system and installation costs, energy costs and water usage over a long period of time (Funt et al., 1980). Main supply lines should be designed for future expanded water usage but are based on the available water supply capacity. Also, growers may wish to install equipment to apply chemicals and fertilizer through the drip system. Fertilizer and chemical injectors are available to add nutrients or other chemicals into the irrigation water for soluble delivery. A dealer can propose the appropriate equipment to meet these needs. Fertigation (see Appendix 2) is an easy way to apply a uniform application of nutrients to the crops in an irrigation zone as required and when needed. Chemicals can be applied per their label.

WEED MANAGEMENT

Raspberry growers need to eliminate weeds to obtain effective plant growth, to reduce moisture and fertilizer loss and to reduce shading and competition among raspberry plants. Weeds that grow tall, have thorns (thistle or horse nettle) or that produce vines that climb over plants impede hand harvesting because people do not want to be injured or they cannot find the fruit. Growers also do not want weeds to restrict mechanical harvest or have pieces of these plants mixed with harvested fruit.

The use of herbicides and cultivation by plowing or rototilling can greatly reduce the weed pressure before, during and after planting. When planting into fields that have been in a monoculture (pasture, hay, or woody plants) for many years, it is wise to use a combination of herbicides and cultivation 1–2 years (years −1 to 0) before planting to reduce perennial weeds and their roots. Using a safe herbicide in the row just after setting the plants can allow much less pressure of weeds in the second or third year. Research indicates that the use of a herbicide in the row and either seasonal cover crops or permanent sod in the drive row (alley) is a system of weed management that can benefit many growers. As described above, using certain grasses to establish a permanent sod can reduce insects in harsh environments. Further, if a thick sod is immediately formed after planting, fewer weeds will emerge, and thus, sod becomes a biological weed control method.

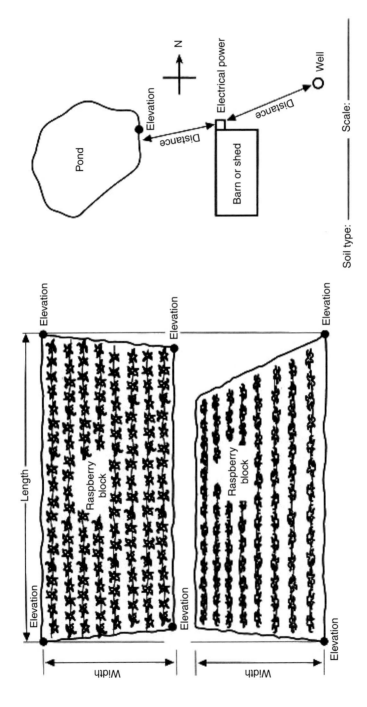

Fig. 9.6. Planning information for irrigation system designer. (Adapted from: Ross, 2004.)

Funt *et al.* (1998) in Ohio applied Oryzalin 4 AS at 4 l/0.4 ha (4 quarts/acre) plus Isoxaben 75 DF at 590 g/0.4 ha (1.3 lb/acre) over the top of transplants several days after planting dormant tissue-cultured black raspberry plants (plugs) and nursery matured plants (plugs grown in the field for 1 year). Transplants were set into cultivated, weed-free soil on raised beds. Straw mulch at 6.4 cm depth (2.5 in) was applied after the herbicide application. This treatment had the highest plant vigor and number of new shoots among the herbicides tested. Isoxaben alone had the fewest dandelion after planting and the following year. Straw mulch improved herbicide effectiveness and soil moisture. In a previous study, Erf and Funt (1984) found that Simazine 80% showed phototoxic symptoms when used on newly transplanted 'Brandywine' purple raspberry.

After planting, a weed-free area about 1–1.5 m (3–4 ft) wide should be maintained in the row middle. Fertilizer and irrigation are more efficiently used when the root zone is maintained free of weeds. Growers may want to apply mulches, such as composted yard waste or sawdust, particularly during the first year. An application of pea straw can be very effective. These can improve water-holding capacity but may contain weed seeds. Root systems tend to be shallower under these systems, and therefore, additional applications in year 2 or 3 may not be advisable. Growers should monitor nitrogen levels and be aware of mice or voles, which tend to nest or burrow in mulches during the autumn or winter. Together pre-emergence herbicides and mulching can prevent the establishment of most types of weeds (Table 9.2). However, some weeds may not be controlled and may need cultivation or careful removal by hand in the first 2 years. The use of two or three different herbicides, either as a single or combined spray and either in the spring and/or late autumn may be necessary in year 4 and later years.

The type of soil can affect the effectiveness of herbicides. Lower rates of some herbicides are used on sandy soils which may contain less than 1% organic matter. Generally, silt loams with more than 2% organic matter may need higher rates to provide seasonal control of weeds and will not reduce raspberry plant growth. In some cases certain herbicides should not be used on certain soil types during the first year. Also, frequent rainfall or irrigation can cause herbicides to be washed out of the top layer of soil and weed control is reduced or weeds are only controlled for one-half of a season. The selection and timing of the application of a herbicide is best made by reading the label, gaining experience on a small piece of land and/or obtaining information from a reputable dealer or consultant.

Table 9.2. Partial listing of herbicides for raspberries, Ohio (Doohan et al., 2010).

Trade name	Common name	Risk of resistance	Signal word	Re-entry interval (h)	Remarks
Pre-emergence control of grasses and/or broadleaf weeds					
Casoron (Norosac)	Flumioxazin	Medium	Caution	12	
Devrinol	Napropamide	Low	Caution	12	
Gallery	Isoxaben	Medium	Caution	12	Non-bearing
Karmex	Diuron	Medium	Caution	12	
Princep	Simazine	Medium	Caution	12	
Sinbar	Terbacil	Medium	Caution	12	
Snapshot	Isoxaben+ Trifluralin	Medium	Caution	12	Non-bearing
Solicam	Norflurazon	Medium	Caution	12	
Surflan	Oryzalin	Low	Caution	12	
Post-emergence control of grasses					
Fusilade	Fluazifop	High	Caution	12	Non-bearing
Poast	Sethoxydim	Low	Warning	12	
Scythe	Pelonic acid	High	Warning	12	
Select	Clethodim	High	Warning	12	Non-bearing
Post-emergence control of broadleaf weeds					
Aim	Carfentrazone	Medium	Caution	12	
Post-emergence control of grasses and broadleaf weeds					
Gramoxone, Inteon Restricted Use Pesticide	Paraquat	Medium	Poison	12	
Roundup	Glyphosate	Low	Caution	12	
Reglone	Diquat	Medium	Caution	24	

REFERENCES

Doohan, D. *et al*. (2010) Midwest Small Fruit and Grape Spray Guide. Bulletin 506B. The Ohio State University, Columbus, Ohio, pp. 1–68.

Erf, J.A. and Funt, R.C. (1984) Effect of herbicides on newly planted 'Brandywine' purple raspberry. *Research Circular – Ohio Agricultural Research and Development Center* 283, 63–65.

Funt, R.C. and Bierman, P. (2000) Composted yard waste improves strawberry soil quality and soil water relations. Proceedings of the XXV International Congress. *Acta Horticulturae* 517, 235–240.

Funt, R.C., Ross, D.S. and Brodie, L. (1980) Economic comparison of trickle and sprinkler irrigation of six fruit crops in Maryland, 1978. Maryland Agricultural Experiment Station Bulletin M950, pp. 1–16.

Funt, R.C., Wall, T.E. and Stokes, B.D. (1998) Effect of new herbicides on tissue cultured black raspberry plants. *Research Circular – Ohio Agricultural Research and Development Center* 299, 07–113.

Lea-Cox, J.D., Ristvey, A.G., Arguedas Rodriguez, F., Ross, D.S., Anhalt, J. and Kantor, G. (2008) A low-cost multihop wireless sensor network, enabling real-time management of environmental data for the greenhouse and nursery industry. *Acta Horticulturae* 801, 523–529.

Maloney, K.E., Wilcox, W.F. and Sandford, J.C. (1993) Raised beds and metalaxyl for controlling Phytophthora root rot of raspberry. *HortScience* 28, 1106–1108.

Ross, D.S. (2004) Drip irrigation and water management. In: Lamont, W.J. (ed.) *Production of Vegetables, Strawberries, and Cut Flowers Using Plasticulture, NRAES-133*. Natural Resource, Agriculture, and Engineering Service (NRAES), Ithaca, New York, pp. 15–35.

Ross, D.S. and Wolf, T.K. (2008) Grapevine water relations and irrigation. In: Wolf, T.K. (ed.) *Wine Grape Production Guide for Eastern North America, NRAES 145*. Natural Resource, Agriculture, and Engineering Service (NRAES), Ithaca, New York, pp. 169–195.

Ross, H. and Auchter, E.C. (1930) *A Production and Economic Survey of the Black Raspberry Industry in Washington County, Maryland*. The University of Maryland Agricultural Experiment Station, College Park, Maryland, pp. 207–245.

Slate, G.L., Braun, A.J. and Mundinger, F.G. (1949) Raspberry growing; culture, disease and insects, Bulletin 719. Cornell Extension Bulletin, Ithaca, New York, pp. 1–68.

10

PRUNING AND TRAINING

RICHARD C. FUNT*

*Department of Horticulture and Crop Science,
The Ohio State University, USA*

INTRODUCTION

Two of the most labor-intensive operations in raspberry production are the pruning and training of the raspberry canes. Yet collectively they have one of the most significant impacts on the overall yield of large fruit and the harvest of disease-free, high-quality berries. In this chapter pruning is broadly defined as removing canes that are no longer producing raspberries, are small in diameter and potentially unproductive; heading of the top of the floricane to cause branching; the removal of broken, diseased or dead parts of the plant; and narrowing of the row for disease control. Training of the canes refers to the use of support systems (trellises) onto which the canes are tied or loosely held between wires for improved ripening and increased yields, for support against strong winds, to improve the efficiency of hand or machine harvest, to prevent heavily loaded fruit-bearing canes from touching the soil, and to reduce disease of fruit and canes.

Most of the commercial and home garden plantings in the USA use a support system or trellis. There are many different types of support systems; some systems are designed for climatic conditions and some for ease of harvest. In the early 20th century, single posts were set by each black and red raspberry plant and plants were tied to the posts (hill system).

In some systems, where only cultivation was used, harrowing (tine cultivator) was done in both directions. Research has shown that planting in rows, using cultivation and equipment only in one direction (hedgerow) is more productive than the hill system. Further, at the end of the 20th century, raised beds were recommended for hedgerow systems, particularly where plants were grown in slow percolating soils. Thus, supported canes, which use a trellis with plants grown in a hedgerow, with or without raised beds,

---

*richardfunt@sbcglobal.net

are described. This chapter will emphasize the various pruning methods and support systems that are being used in commercial plantings around the world.

PRUNING AND TRAINING

Red raspberries – summer-bearing floricanes

During the planting year, summer-bearing red raspberries grow upright and produce fruit buds in late summer and autumn. Generally, no pruning is necessary in the first year unless there are injured or dead canes caused by wind or poor management. In the second and subsequent years, floricanes can be pruned in late winter to early spring before bud swell. These floricanes will produce leaves and fruit in late spring and early summer and are referred to as summer-bearing types.

In mild climates (hardiness zone 8), with low winter temperatures of −7 to −12°C (+20 to +10°F), floricanes can be removed as close to the soil as possible in late autumn, winter or early spring. Plants should not be pruned after harvest. Research has shown that nutrients and carbohydrates move from the dying tissue of the canes into the crown and roots. Delaying cane removal until the plants are completely dormant results in a higher level of winter survival and better plant growth. However, if canes are damaged and are showing disease, such as yellow rust, then they should be removed immediately after harvest.

In cold climates (hardiness zone 6 or 7), with low winter temperatures of −12 to −23°C (+10 to −10°F), spent floricanes are removed in late winter to early spring. Where deep snow occurs, the spent (old) canes are left until spring and as soon as the snow melts, the canes are cut near the soil and moved to the side of the row. After the lowest temperature of winter has passed, winter-damaged canes are removed along with other unwanted canes. Canes that are short and/or small in diameter (less than pencil-sized or 0.6 cm (¼ in) diameter) are removed at the base of the plant. Plants that have grown outside of the 0.3–0.45 m (12–18 in) row width should be removed (Fig. 10.1). Do not allow a row width of more than 0.61 m (24 in) during the growing season so that air movement and disease and weed control (herbicide coverage of soil) aren't reduced (Funt *et al.*, 1999). A row width of 0.3–0.45 m (12–18 in) is considered ideal for harvesting. When cane removal is completed, only four to six canes per 0.3 m (1 linear ft) or 24 to 36 canes per square meter (or approximately 1 square yard) remain. These canes have the largest diameter at the base and generally are the tallest canes, ranging in height from 1.2 m to 1.8 m (4–6 ft) and up to 2.7 m (9 ft).

Floricanes are headed (tipped) by cutting a portion of the top of the canes that remain. Again in cold climates, this should be done after the lowest

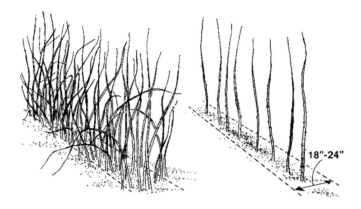

Fig. 10.1. Red raspberries planted in a narrow hedgerow. Plants growing outside the row are cut off at ground level. (Courtesy of The Ohio State University.)

temperatures have occurred. All winter-damaged canes should be cut below the damaged area. Vigorous canes can be left longer than shorter ones. On light soils and under dry conditions, the canes may be cut back more severely than on irrigated soils or soils having abundant rainfall. However, removing more than 0.1–0.15 m (4–6 in) can reduce yields (Fig. 10.2). However, in the interest of reducing labor, mechanical pruning (heading) can be done and all plants are left at the same height. Cane height varies with cultivar and growing conditions. If the cane height is greater than 1.5 m (5 ft), hand harvest may be difficult. The tops of plants that are 2–2.7 m (6–9 ft) in height can be cut back as described or bent downward and tied onto a support system for either hand or mechanical harvest. When canes are shorter than 1.2 m (4 ft) they may be cut back (below the damaged area) when winter damage or powdery mildew is prevalent. Removal of the top is important for powdery mildew control.

The cultivar 'Prelude' will generally produce a lot of fruit on the top 0.15–0.2 m (6–8 in) as a primocane in the previous autumn and then produces as a floricane in early summer. The area where autumn production occurred is removed during the dormant pruning. Generally, this leaves 1.2–1.5 m (4–5 ft) of plant for summer production.

Summer-bearing red raspberry primocanes are not headed during the growing season. Heading would increase their vigor and cause an overly dense canopy, reducing disease control. The spent canes are first removed from the row in the dormant season. The former primocane is now a floricane and is allowed to remain. This is called selective pruning and has been the most favored method worldwide during the 20th century. Floricanes are thinned by cutting at ground level as described above. This is done entirely by hand and no mower or suppression of canes is used. Canes are then gathered and

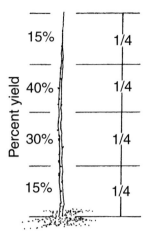

Fig. 10.2. Percent yield of red raspberry plant. Do not prune back more than one-quarter of height from the top or yield will be reduced. (Courtesy of The Ohio State University.)

burned or chopped by a flail or rotary mower so that they can easily decay, thus reducing disease pressure.

Primocane suppression, partial primocane suppression, alternate-year mowing and alternate-year mowing with primocane suppression are methods that either increase berry size and/or reduce labor and fungicide sprays. These systems should start in the third year and not be used on weak sections of the field. While there are some advantages to these systems, there are also disadvantages (Bushway et al., 2008).

Training-support systems

In commercial plantings, red raspberries can be grown using individual stakes (hill system, Fig. 10.3a) or as an unsupported hedgerow (Fig. 10.3b). However, a trellis (hedgerow system) consisting of many wood (treated or non-treated) or metal 'T' posts and either metal or plastic (monofilament) wire or string made of plastic, jute or hemp (Fig. 10.3c and d) is generally used in commercial plantings. A support of either individual stake or trellis hedgerow type is used to keep the large number of berries and fruit weight on the single canes from touching the soil during harvest. The hill system has been widely used for cultivation in both directions for weed control by harrowing, rototilling or hand weeding. In Washington and Oregon, hill trained raspberries refers to groups of canes on a four-wire trellis (top wire at 2 m/5–6 ft) or hedgerow trained where canes are individuals. Plants can

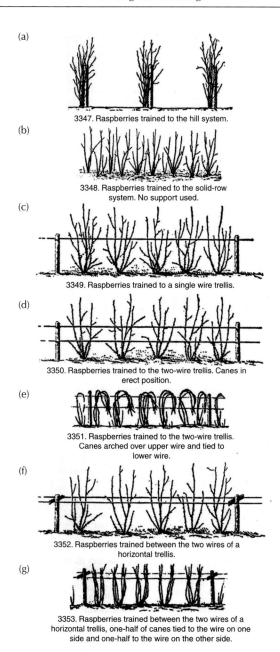

Fig. 10.3. (a) Hill system. **(b)** Solid row system. **(c)** Planted to a single row system. **(d)** Raspberries planted to a two wire system. **(e)** Plants trained between two wires. **(f)** Plants on two wire system and arched over top wire and tied to bottom wire. **(g)** One half of plants tied to one side and half tied to the other side. (With permission from L.H. Bailey, The MacMillan Company, 1943.)

be on raised beds or flat ground at 14–45 cm (12–18 in) wide rows. In New Zealand, hedgerow plantings have plants spaced at about 30 cm (12 in) in the row. In Oregon, 15 canes per hill out-produced five to ten canes per hill (Barney and Miles, 2007). However, susceptibility to disease increases as plant density increases, reducing returns (income). Further, the time needed to separate the good from the bad fruit was labor intensive and frustrating. From a management point of view, fresh raspberries need to be free of soil and mold, firm with good size, and shiny for a high level of consumer satisfaction.

A hedgerow system does not allow any movement of people or machines across the row. Every operation moves down the row on either side of a trellis that is made of posts set every 6–8 m (20–28 ft) with two, four or six wires or string. Generally, hedgerows are more productive per unit of land than the hill system and accommodate tractors (1.5 m/60 in wide) and implements for insect, disease and weed control, and mechanical harvest. For canes taller than 1.2–1.5 m (4–5 ft) the top wire should be at 0.9 m (3 ft) or as a rule placed at least two-thirds of the plant height (Fig. 10.4).

There are several types of trellises used around the world including the I-trellis, which uses a single wire (Fig. 10.3c) or two wires with one set above the other (Fig. 10.3d and 10.3e). The V-trellis has posts that are set at a 20° to 30° angle with three wires on each side; the top wires are 1.2 m (3.5 ft) apart and fruiting canes tied to one side and non-fruiting canes tied to another side using the three wires. In the USA, a T-trellis has either two wires (0.5 m/2 ft) above the ground with cross arms about 0.6 m (2 ft) across (Fig. 10.3f and 10.3g) or two sets (double T) of wires and cross arms at 0.5 m (2 ft) and 1 m (4 ft) in height from the ground with cross arms at each post that are 0.5–0.6 m (18–24 in) long, respectively. The length of the cross arms maintains row width and height of canes. Canes are tied to either side onto the top wire only. Tying the canes to the wires at 1.2 m (4 ft) allows easy access to the fruit at harvest and the width of 0.6 m (2 ft) provides primocanes with sunlight to grow tall in the center of the row. Wires are permanent and are not moved.

The Gerde system from Norway can be used with a T-trellis having one set of two wires strung across the top of the cross arms. Wires are set into grooves

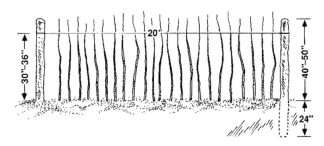

Fig. 10.4. Permanent trellis posts are spaced about 6 m (20 ft) apart. Set posts in the ground to about 0.6 m (2 ft) into the ground. (Courtesy of The Ohio State University.)

that have been cut into the cross arms to make a narrow width for dormant canes, and as flowering occurs, the wires are moved to an outer notch in the wooden cross arms and then primocanes are allowed to come up in the center of the row. This movement makes harvesting more visible and efficient. However, care must be taken not to damage the canes when they are moved (Pritts and Handley, 1989). Trellis end and line posts should be made from materials that will last for the potential 15-year life of the planting.

Red or yellow autumn-bearing primocanes

In the USA, red or yellow autumn-bearing primocanes, grown in the field for one crop per year, are pruned (cut) at 2.5–5 cm (1–2 in) above the ground, either after harvest in warm climates or in the spring before bud break in cold climates (Barney and Miles, 2007). In humid climates, a dormant spray is applied prior to cane removal to reduce the pressure of cane diseases. Where a trellis support system (single or double T) is used, temporary posts and string or plastic wire are removed and the canes are cut with a mower or special hand-held gas-powered trimmer.

In the late spring or early summer, it is wise to narrow the row, particularly several weeks before bloom. This allows plants to grow taller and more water and nutrients to reach those canes between the wires. Plants need to be of a height, with narrow rows, for ease of harvest by pickers (Plate 10.1). Rows are narrowed to 45–60 cm (18–24 in). In this system, primocanes are not thinned within the rows to manage cane density. In climates having a long, frost-free autumn, careful removal of the tip (heading or tipping) above an auxiliary bud can increase fruit production. For field-grown primocane raspberries, heading is not suggested for the Midwest or Pacific Northwest due to a short autumn season. However, primocanes are sometimes headed during the summer to delay fruit set for off-season production in greenhouses or hoop houses where harvest is extended into late autumn.

Black and purple raspberries

In the first planting year (year 1) the 'handle' or main stem is cut and removed several weeks after planting to reduce disease spreading from the stem onto the new soft stems. If the grower receives clean plants and applies a fungicide during the season, this practice may be unnecessary. No other pruning is suggested during the first year.

Plants that are subjected to good weed control and frequent irrigation during the first year can, in the second year, produce stems and many laterals that are 1–1.2 m (3–4 ft). These stems and laterals can be detached from the soil at the tip and single canes are tied to the trellis wire (support system) and then headed in a manner to extend 20 cm (8 in) beyond the wire for easy picking (Fig. 10.3g). In some cases, laterals are thinned to maintain some space for air movement or canes that are too small in diameter are removed. Moderate yields (first baby crop) can be harvested. If there are few stems or laterals to reach the wire, it may be best to cut most of the growth and leave an untied stem 30 cm (12 in) long. This will allow sunlight into the interior part of the row for the primocane to grow.

Before harvest in the second or subsequent years, primocanes (new canes) are headed at 60 cm (24 in) for stocky and self-supporting canes (no support system) (Slate *et al.*, 1949). This is referred to as raspberries trained to the solid row system (Fig. 10.3b). Canes can also be headed at 90 cm (36 in) for a T-trellis system. In either case, allow the canes to grow 10 cm (4 in) above the recommended height and remove this part from the top, leaving the cane 60 or 90 cm (24 or 36 in) tall (Plate 10.2 and Fig. 10.5). The heading (tipping) can be done by hand pinching with gloved hands, with a sharp knife, or with pruning shears. Because all plants do not reach the same height at the same time, the pruning may need to be done every few days. A fungicide spray, within 24 h after pruning, on this soft wood will reduce diseases entering the cut. If the plants are not pruned and grow above a height of 1.2 m (4 ft), the laterals will be short and stems will be weak. No additional pruning of canes or laterals should be done until the canes are dormant or after the coldest temperatures of the winter have been reached.

Generally, the heading at 90 cm (36 in) reduces stooping to pick the berries from laterals that grow between 90 cm and 1.2 m (36 in and 48 in). Therefore, people can stand erect and reduce spillage from either their hands

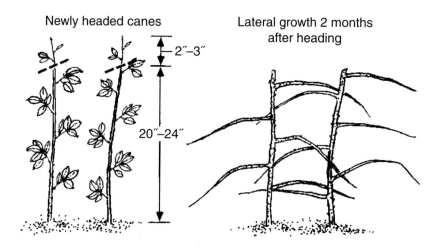

Fig. 10.5. Heading (cutting back from tip) of black or purple raspberry in summer (left); lateral growth after heading (right). (Courtesy of The Ohio State University.)

or container. It appears that these taller canes may be better for mechanical harvest. They are also easier to prune. However, primocanes headed at 60 cm (24 in) can be self-supporting, requiring no trellis, and therefore, reduce establishment costs.

After the floricanes have fruited, the spent (old) canes should be removed by cutting the stem near the soil. In warm climates (zones 7 to 8), spent canes can be removed in early to late autumn without subjecting the canes to possible winter injury or to delayed acclimation. In cold climates (zones 5 to 6), canes should be removed in late winter or early spring.

Prior to the removal of the spent black raspberry canes, laterals, which can be 90 cm (3 ft) or longer are cut back to 20–35 cm (8–14 in) in length (Fig. 10.6), leaving eight to 12 buds (Slate *et al.*, 1949). Cutting back the laterals prior to the removal of spent canes makes it easier to remove spent canes to the aisle (Plate 10.3). Most of the fruit buds are on the plant's laterals; laterals that are pencil sized in diameter (0.6 cm/0.25 in) can be pruned to a longer length than those that are smaller in diameter. Fruit size will be larger on thick laterals (Plate 10.3). Laterals smaller in diameter than a pencil should be pruned at 15–20 cm (6–8 in) to improve fruit size. The pruning should be completed after the severest cold weather has passed so as to be able to remove canes or laterals that have been damaged by low winter temperatures.

Floricanes (new canes) are thinned to four to six canes located in clusters of the original plant. Remove canes that are less than 1.2 cm (½ in) in diameter at the base of the plant. A large number of canes per unit of land (14,000 to 16,000 per hectare (5500 to 6500 per acre)) are preferred

Fig. 10.6. Black and purple raspberry before (left) and after (right) dormant pruning. (Courtesy of The Ohio State University.)

as compared to fewer canes and longer laterals. The plants remaining are oriented to both sides of the row, with each plant and lateral independent (no crossing over) of each other, so that the berries and canes are receiving good air movement and are easily harvested. With a V-shape from the top to bottom, sunlight can reach the soil and allow strong plants to grow for the next year. Flower bud initiation occurs in late autumn, followed by flower development in late spring. The newly developed spineless black raspberries may be similar in growth habit to red raspberries and could be managed in a similar way.

Purple raspberries

Purple raspberries are crosses between red and black raspberries which produce reddish to purple fruit. Many purple raspberries are pruned similarly to red raspberries and are not headed in the summer. Other cultivars of purple raspberries, such as 'Brandywine', will respond to summer heading like black raspberries. Heading should occur at 90 cm (3 ft) on these types. Generally, purple raspberry cultivars in the USA are vigorous and grow very tall. Leaving three to four fruiting canes per linear foot (30 cm) of row is acceptable. Both types will require some type of trellis. Primocane suppression can be used to control this vigor with good results (Pritts and Handley, 1989).

MECHANICAL HARVEST

Training systems that are unsupported do not do well for mechanical harvest. In most cases the canes bend over and are not erect, and thus difficult to pick by machine. Supported two- or four-wire systems have shown good results (Fig. 10.3d and e) in the removal of fruit over two or three harvests. Generally, the mechanical harvester in the USA has been designed to accept a 10-cm (4-in) diameter post and 2–3 m (7–10 ft) in height without damaging the machine (Plates 10.4 and 10.5).

HIGH TUNNELS AND GREENHOUSES

In most cases floricane and primocane red raspberries have been tested in high tunnels and greenhouses mainly due to their wide market acceptability. They are more adaptable and will produce more fruit per cubic meter than black raspberries. Economically, large volumes of fruit in the 'off' season (normal local field production) gain a higher price, which offsets the higher investment. Growers in the eastern USA indicate that raspberries are profitable if the structure does not need additional light and heat. In recent years, more raspberries and strawberries are being brought from fields in other regions and competing with local out-of-season berries.

An I-trellis with wood end posts and metal 'T' posts in the row and short cross arms are recommended in high tunnels. In greenhouses with concrete floors, bamboo posts are set in a bucket of sand and the top is tied to the greenhouse structure (Pritts and Handley, 1989). Trellis systems are from 1.8 m to 2 m (6–7 ft) in height and narrow rows are more productive than wide rows.

Pruning begins the first year because emerging plants can easily exceed the top of the trellis. Thin out the smallest and largest canes at ground level and leave four to six healthy primocanes per linear foot in late summer. Head these plants in the dormant season by cutting back the tops by 10–20 cm (4–8 in). These will fruit in early spring. In the spring, remove (cut to the ground) the first flush of canes and allow the second flush to grow. As the primocanes reach 1–1.5 m (3–4 ft) they should be headed; this will slow down their growth and allow more light to the flowers and fruit. As the primocanes break bud, new upright growth will occur below this cut and the stems will become larger in diameter. Several headings may be necessary under ideal conditions (Pritts and Handley, 1989). Remove the spent canes in early fall.

Autumn-fruiting primocane cultivars will begin to produce flowers in late summer after an early spring planting. After a few berries have been harvested, head (tip or pinch) the top 5–10 cm (2–4 in) of the plant (spent portion) to encourage branching and additional fruiting over an extended period of time beyond the first frost. Some autumn-bearing cultivars will naturally branch at this height and may not need heading. New primocanes will emerge about every 2–3 weeks during the season, and therefore require heading. For a delayed harvest in high tunnels, head the plants before flowering when plants are 0.5–1 m (2–3 ft) tall. The earlier the cut, the greater the delay in harvest will be. This is especially good for late fall harvest, which captures a low supply in the market (Pritts and Handley, 1989).

REFERENCES

Bailey, L.H. (1943) *The Standard Cyclopedia of Horticulture*, Vol. 3. The Macmillan Company, New York.

Barney, D.L. and Miles, C. (eds) (2007) *Commercial Red Raspberry Production in the Pacific Northwest*. PNW 598. A Pacific Northwest Extension publication.

Bushway, L., Pritts, M. and Handley, D. (eds) (2008) *Raspberry and Blackberry Production Guide (NRAES-35)*. Natural Resource, Agriculture and Engineering Service (NRAES), Ithaca, New York.

Funt, R.C., Bartels, S., Bartholomew, H., Ellis, M., Nameth, S.T., Overmyer, R.L., Schneider, H., Twarogowski, W.J. and Williams, R.N. (1999) *Brambles: Production, Management and Marketing*. Bulletin 782. The Ohio State University Extension, Columbus, Ohio.

Pritts, M. and Handley, D. (1989) Pruning and trellising brambles. In: *Bramble Production Guide (NRAES-35)*. Northeast Regional Agricultural Engineering Service, Ithaca, New York, pp. 41–51.

Slate, G.L., Braun, A.J. and Mundinger, F.G. (1949) *Raspberry Growing – Culture, Disease, and Insects* (revised edition). Cornell Extension Bulletin 719. New York Agricultural Experiment Station, Geneva and Cornell University, Ithaca, New York.

11

PEST AND DISEASE MANAGEMENT

RICHARD C. FUNT*

*Department of Horticulture and Crop Science,
The Ohio State University, USA*

INTRODUCTION

In this chapter, the effects and control of insects and other pests, diseases and weeds in raspberry production are discussed. The primary effect of weeds in raspberries is the competition for water, nutrients and sunlight. However, weeds can harbor disease, nematodes and insects, and their management and control is vital for successful production and marketing of raspberries. Weed control is also discussed in Chapters 9 and 12. Application equipment (sprayer technology) for insect, disease and weed control is also discussed in this chapter.

Insects transmit certain diseases in raspberries, particularly viruses, and can cause damage to roots, canes and fruit. Infestations of insects are generally correlated to the local climate, such as periods of warm, windy conditions. Therefore, insect control can reduce disease pressure. Semi-arid regions may have fewer insect species or have areas with predators, and thus fewer insects. However, greater numbers of nematodes can be found in fields that have had the same crop (monoculture) for many years. Therefore, planting raspberries in such soils may cause an increase in the number of certain diseases or the spread of certain viruses. In these cases, it is wise to maintain certain pests below a damaging threshold before planting by using a nematicide or fumigation and maintaining low levels of nematode damage in the planting by means of daily irrigation or chemical injection within the irrigation system.

Diseases, including those caused by viruses, fungi and bacteria, vary greatly among raspberries or within the species of raspberries and/or between humid or semi-arid regions and between soil types (sandy versus clay), which tend to be dry or wet, respectively. Excessive or deficient levels of plant nutrients can affect disease levels, together with high temperatures, chemical toxicity,

*richardfunt@sbcglobal.net

hail and wind. Research in the USA has identified certain cultural practices (raised beds, compost, pruning) for reducing root rot, increasing air movement (site selection) or maintaining good air movement around plants (thinning and pruning of canes) to provide drying of stems, leaves and fruit for reducing fungi development, avoiding disease-infected soils and reducing the use of overhead irrigation which reduces disease pressure (see Chapters 9 and 12).

Research in the USA in the last half of the 20th century has identified new and more effective fungicides and bactericides. Further, there has been an increased use of application equipment (sprayers) for fungal disease control (manual, high pressure or motor driven air blast types) that provide rapid applications and improved results in obtaining complete coverage of the canopy and fruit. Research and outreach programs in integrated pest management (IPM) in which US growers monitor for insect, disease pressure, and weeds have shown good preventative strategies and have reduced the number of chemical sprays compared to conventional methods. Reducing the number of trips made by machines through the field and reducing the units of chemicals at the same time improves plant health and fruit quality, allowing for a higher return on the investment. Other means of controlling pests comes from the purchase of essentially virus-free plants, genetic resistance, cultural practices (removal of spent plants by burial, burning or chopping), maintaining plant vigor, good berry floor management and complete picking of ripe berries at harvest.

DISEASES OF ROOTS

Verticillium wilt

Verticillium albo-atrum Reinke and Berth and *Verticillium dahalie* Kieb is a serious disease, particularly of black and red raspberries in the USA and Europe (Jennings, 1988). Since 2000, the incidence of Verticillium wilt in the UK has been more common and significant (H. Hall, n.d., personal communication). The fungus is long lived in soils and persists in an active growing state or as a dormant resting structure. Infection occurs when raspberry roots come into contact with the active fungus, which goes into the xylem tissue and causes the plant to wilt, particularly as mid-summer temperatures cause stress in the plant (Plate 11.1 and Plate 5.12).

The level of *Verticillium* severity is dependent on the type of plants (solanaceous types as tomato, potato, tobacco, etc.) or weeds (pigweed, nightshade, horse nettle and lambs quarter) that were present before raspberries were planted. These fields can be fumigated or planted with non-host plants (small grains, corn) for 4–5 years before planting raspberries. While beneficial, this may not solve the problem entirely. Moving to or purchasing a field that has not had such plants previously to planting plants purchased from a reputable nursery and/or purchasing red raspberry cultivars having some resistance to a few strains of the fungus is a better management strategy (Bushway *et al.*, 2008). Black and purple raspberries are more susceptible than red raspberries. *Verticillium* is a cool weather disease and is most severe in poorly drained soils following cool, wet springs. Plants show signs of wilting in early to mid-summer. The leaves of black raspberries at the base of the plant may appear as a dull green, compared to the brighter green of normal leaves. A bluish streak can be seen on the cane (Plate 11.2). Starting at the base and moving upward, the leaves wilt, turn yellow and drop away from the plant (Funt *et al.*, 2004).

Phytophthora root rot

Phytophthora fragariae var. *rubi* (Wilcox and J.M. Duncan) W.A. Man in't Veld (2007) is the most serious disease afflicting red and purple raspberry roots around the world. It is found everywhere raspberries are grown, except in New Zealand (although other species of *Phytophthora* are found in New Zealand). This disease is found particularly in wet soil conditions or where there is abundant rainfall or a good supply of water, as with irrigation. Black raspberries are much less susceptible to the disease. Several root rot diseases have been associated with infections of the genus *Phytophthora*, such as *P. megasperma* in the UK and the USA; *P. crytogea*, *P. citriocola* and *P. cactorum* have been found in North America and Europe, particularly in heavy soils that tend to be wet. Infected plants become weak, stunted and are susceptible to winter injury; they will collapse and die (Plate 5.8).

The first symptoms appear in mid-summer to early fall; healthy plants will lack vigor and show symptoms of 'wet feet'. Leaves become yellow, red or orange and begin scorching (browning) on the edges as the plant wilts and dies (Plate 11.3). Because other symptoms look similar to *Phytophthora*, plants should be dug up and the outer layer scraped off the main roots and crown. Healthy plants have white roots and crown; infected plants will be red and turn dark as the tissue decays. Generally, a distinct line can be seen between the unhealthy root (below ground portion) and the healthy part of the plant (Plate 11.4).

In wet soils, the fungi persist and develop to where the zoospores are formed and expelled into the soil. As the soil becomes saturated, it has little oxygen or pore space (Funt *et al.*, 2004). The zoospores attach themselves to the root and begin their infective process. Within an established raspberry field, management of Phytophthora root rot can be cultural and/or chemical. If possible, a root rot resistant cultivar should be used, so that problems with this disease are minimized or eliminated.

All raspberries should be planted on well-drained soils. Complete a percolation test by digging a hole to 35 cm (14 in), pour in 20 l (5 gal) of water and return in 1 h. If water remains in the hole it is not considered to be well drained. If planting in this soil, growers should refrain from susceptible cultivars and plant quality nursery stock. A reduction in disease severity can be provided by raised beds where the soil water tensions in the root zone exceed zoospore activity (Maloney et al., 1993). However, the combination of metalaxyl fungicide (Ridomil is a common name) plus raised beds provided optimal control for all measures of disease incidence and severity. The use of raised beds plus compost incorporated into the top 10 cm (4 in) can lower water tension and increase porosity, and may also be of benefit in reducing the severity of root rot (see Chapter 9). Because raised beds are drier than a flat surface, drip irrigation should be used when rainfall is insufficient for growth. Development of resistant cultivars in North America has been ongoing in Oregon, Washington State, New York and British Columbia. As a result of severe problems with Phytophthora root rot in the UK, Europe, Scandinavia and Australia, breeding programs have produced resistant selections and/or cultivars in each of these areas.

Bacterial crown gall

Bacterial crown gall (*Agrobacterium tumefaciens*) and cane gall (*Agrobacterium rubi*) are widespread diseases of all raspberries and affect black and purple raspberries more than red raspberries (Plate 11.5). These are particularly serious in nursery fields and total control is essential for preventing the spread of the disease. Galls form each season and vary in size from the size of a pin head to several centimeters (inches) in diameter. They form on fruiting canes, are spongy, wart-like tumors, and become hard brown to black woody knots (Funt et al., 2004). These galls can be found on the roots, crowns or canes and can interfere with water and nutrient flow.

Crown galls enter the plant only through natural openings, winter injury damage, soil insects or nematodes. Root pruning or cultivation increases the incidence after planting, and because the bacteria survive in the soil for many years, the roots are infected. Further, as the bacteria grows, the plant sloughs off bacteria, which overwinters in both galls and soil; in the spring the bacteria are spread by splashing rain, irrigation, pruning tools and insect feeding (Funt et al., 2004). The disease is most severe in cool, wet soils having a high pH (Jennings, 1988). The disease can be managed by planting healthy plants, avoiding infected fields, avoiding close cultivation and choosing resistant cultivars, but these are very limited. Resistance to the disease, as in the cultivar 'Willamette', is significant, but no breeding efforts, as of 2011, have been put into selecting for resistance to this disease. Although cultivars derived from 'Creston' and 'Skeena' are susceptible to *Agrobacterium* infection, these cultivars are still used for breeding.

DISEASES CAUSED BY VIRUSES

Nearly 26 viruses have been reported for *Rubus* crops around the world and are classified into four groups, by their mode of transmission: (i) by large aphids (*Amphorophora* species); (ii) by small aphids (*Aphis* species); (iii) by nematode species; and (iv) by infected pollen. Additionally, *Rubus* stunt is associated with a mycoplasma-like organism. *Rubus* stunt is an important disease in Russia (former Soviet Union) and Eastern Europe, but is rare in Western Europe and does not occur in North America. Raspberries suffer more from viruses than any other fruit crop in the USA and can reduce yields by at least 70% (Funt *et al.*, 2004). Greenhouse or laboratory tests using black raspberries as indicators (refer to Plate 6.3) are needed for positive identification of viruses because there are several disorders such as nutrient deficiencies, powdery mildew, pesticide injury (see Chapter 12) and/or feeding insects and mites that cause similar leaf symptoms to viruses (see also Plate 6.5, for tomato black ring virus (TBRV)).

Viruses (also referred to here as mosaics (see Plate 5.11)) are transmitted by aphids, which acquire them after they feed on infected plants for an hour or less and transmit them to healthy raspberries without a latent phase in about 15-min feed. After feeding on healthy plants the vectors lose their virus charge because there is no multiplication of viruses in the vector. *Amphorophora idaei* and *Amphorophora agathonica* are the main vectors for raspberries in Europe and North America. Raspberry mosaic has been introduced into Australia and Chile by infected transplants, but the vectors do not occur there to spread the virus. In Europe and North America breeding programs have been for resistance to the aphid vectors (Jennings, 1988).

Rubus yellow net virus

Rubus yellow net virus (RYNV) and black raspberry necrosis virus (BRNV), found in Europe and North America, can be caused by a combination or multiple infections in certain cultivars. It is difficult to diagnose the presence of these viruses when they are present together. Such mixtures can cause severe reductions in cane numbers, height and fruit size, form, and structure. Raspberry leaf curl virus (RLCV) was first noticed in New York in 1880 and was one of the first red raspberry disorders to spread from plant to plant. The disease is found in red raspberries in New England and the Rocky Mountains and in black raspberries in Michigan and Ohio (Jennings, 1988).

Raspberry bushy dwarf virus

Raspberry bushy dwarf virus (RBDV) is found in Europe, the UK, the USA, Canada and New Zealand. It frequently causes poor fruit set in red raspberries; these fruit are crumbly when they ripen. It reduces vegetative growth in both red and black raspberries. Some red raspberry cultivars contain the *Bu* gene and are immune to the common strain of RBDV (Jennings, 1988). It is very common where susceptible red raspberries are grown and is transmitted by seeds and pollen. When transmitted by pollen, it spreads rapidly once plants begin to flower, reducing the life of the planting to 5 or 6 years, as in 'Meeker' in Washington State and British Columbia, rather than a full life of 12 or more years. Significant variation in the ribonucleic acid (RNA) of RBDV is known. This variation may be the cause of different rates of infection in different environments (Jones *et al.*, 2000). A variant of RBDV imported to East Malling Research (EMR) in the UK in infected seed from Russia became known as the 'resistant-breaking strain' of RBDV. This variant was found to infect many cultivars containing the *Bu* gene and the only cultivars showing some resistance are 'Haida' and 'Schonnemann'.

Raspberry ringspot virus

Raspberry ringspot virus (RRV) or tomato ringspot virus (ToRSV) is found in Scotland, northwestern and northeastern USA, and in other major red raspberry production areas. It causes 'crumbly berry' (see Plate 6.3). Infection with this virus can result in high economic losses due to the production of poor quality fruit and stunted plants. The symptoms show up on the leaves in the spring after infection as yellow rings or fine yellow lines along the veins (Bushway *et al.*, 2008). Plants can be infected in circular patterns, where hotspots of nematodes colonize. This virus is transmitted by nematodes, such as the *Longidorus* and *Xiphinema* species (dagger nematode), where viruses can be retained for several months under ideal conditions. The virus is spread by plants and weeds, such as dandelions, fruit trees (apple and peach), and grapes, particularly where monocropping has occurred over many years. Small crumbly berries are produced and plants are less vigorous because the nematodes infect the roots.

Control of virus diseases begins by starting with certified disease-free and essentially virus-free (virus indexed) plants (see Chapter 6). The most effective way is to start with healthy 'mother' stock plants. Such schemes are found in many countries for the propagation and certification of virus-tested plants. In many cases, this is determined by negative results in graft tests to *Rubus henryi* and/or black raspberry indicator plants. Start several years ahead of planting by destroying plants that are hosts to diseases, insects and nematodes in the field to be planted. Plant grain crops, such as corn, wheat or rye, which are not themselves host crops or reduce weeds and plants that carry disease. Remove infected plants and roots from the producing field. Destroy all wild bramble plants within 180–330 m (600–1000 ft), do not plant black or purple raspberries near old plantings of red raspberries, which may have latent infections, or plant black raspberries upwind from red raspberries. Remove infected plants from the planting by digging and destroying infected plants, and control aphids, particularly in late spring and early summer. Control nematodes before planting and monitor parasitic nematodes with soil testing taken in mid-summer when soil temperatures are above 10°C (50°F) (Table 11.1).

LEAF, CANE AND FRUIT DISEASES

Orange rust

Orange rust (*Arthuriomyces peckianus* (E. Howe) Cummins and Y. Hiratsuka and *Gymnoconia nitens* (Schwein.) F. Kern and H.W. Thurston), important only to North America, attacks black raspberries but not red raspberries (Jennings, 1988). It attacks the leaves as they unfold in the spring (Plates 11.6 and 11.7). Black spores are produced 21 to 40 days after infection and are difficult to see on the underside of the leaf. Fungicide sprays are necessary from late summer into the early fall to reduce the spread of the disease. Infected plants produce orange rust pustules each spring, and because the disease is systemic, these orange leaves appear each spring for as long as the plant is alive. In mid- to late spring, the wind and perhaps splashing rain spread the bright orange aeciospores from the pustules to infect healthy leaves of uninfected plants (Plate 11.6). Orange rust is favored by low temperatures and high humidity. The fungus over-winters as a systemic, perennial mycelium within the host. It is best to identify these plants and remove them from the planting and remove or destroy any plants in the nearby tree lines or neighboring plantings that show signs of orange rust. Partial control of the fungus can be achieved with judicious fungicide sprays as the fungus appears in the spring and again in late summer to early fall when the temperatures are favorable for the development of the disease (Funt *et al.*, 2004).

Late leaf rust

Late leaf rust (*Puccinastrum americanum* (Farl.) Arth and *P. arcticum* Transzschelis) is a disease that only affects red and purple raspberries, develops in late summer and has been diagnosed in most raspberry production regions of the USA and Canada. Heavy losses start with the green fruit and damage 30–100% of marketable fruit on autumn-bearing primocane cultivars, such as 'Heritage' and other susceptible cultivars, such as 'Festival'

Table 11.1. Bramble disease control strategies. (Source: Funt et al., 2004.)

Disease control	Viruses[a]	Verticillium wilt	Orange rust	Cane blights[b]	Powdery mildew	Fruit rot
Good air/water drainage	–	–	–	++	+	++
500+ ft from wild brambles	++	–	–	–	–	–
Rotation	+[c]	++[d]	–	–	–	–
Cultivar tolerance/resistance	++[e]	++[f]	++[g]	–	+	–
Avoid adjacent plantings	++[j]	–	++	–	+	–
Eliminate wild brambles	++	–	++	–	+	–
Disease-free stock	++	++	++	++	+	–
Aphid control (vectors)	++	–	–	–	–	–
Rogue infected plants	++	–	++	–	–	–
Speed drying (weeds, pruning)	–	–	++	++	–	++
Prune 3 days before rain	–	–	–	++	–	–
Dispose of diseased pruned canes	–	+	+	++	–	–
Maintain plant vigor	–	–	–	++	–	–
Fungicide sprays	–	–	++	++[h]	++[i]	–
Harvest before over-ripe	–	–	–	–	–	++
Fruit storage conditions	–	–	–	–	–	++

Key: ++, most important controls; +, helpful controls; –, no effect.
[a]Viruses: mosaic (raspberry), leaf curl (raspberry, with blackberry symptomless), ringspot (red raspberry) and streak (purple and black raspberry).
[b]Cane blights: anthracnose, cane blight spur blight and Botrytis blight.
[c]Rotation effective for ringspot virus only; 2 years of grass crop (e.g. corn) with excellent weed control before planting red raspberry should eliminate need to fumigate for *Xiphinema*, a nematode vector.
[d]Rotation for Verticillium wilt: avoid fields planted to susceptible crops (tomatoes, potatoes, eggplant, peppers, strawberries, raspberries, stone fruit) within the past 5 years. Avoid fields with history of Verticillium wilt unless soil is fumigated.
[e]Virus resistance, tolerance and immunity: mosaic – black and purple raspberries are more severely affected than red raspberries. Of purple and black raspberries, 'New Logan', 'Bristol' and 'Black Hawk' are tolerant; 'Cumberland' is susceptible. Of red raspberries, 'Milton', 'September', 'Canby' and 'Indian Summer' are 'resistant' because aphid vectors avoid them. Leaf curl – all raspberries are affected. Tomato ringspot – red raspberries are affected. Streak – black and purple raspberries are affected.
[f]*Verticillium* tolerance: most black berries are resistant; red raspberries are more tolerant than black raspberries. 'Cuthbert' and 'Syracuse' red raspberries appear to be resistant under field conditions.
[g]Orange rust resistance: red raspberries are immune. Other brambles are affected.
[h]Fungicide program for cane blights: the lime-sulfur spray (delayed dormant) is most important for anthracnose and cane blight.
[i]Fungicide program for powdery mildew: sulfur will provide good control of powdery mildew.
[j]Keep black and purple raspberries away from red raspberries because mosaic virus can spread from red raspberries and is more severe on black and purple raspberries.

and 'Nova'. The small, numerous yellow spots are seen on the underside of the leaf. The fungus probably over-winters on raspberry canes. Fungicide sprays are recommended to overcome the damage from this disease (Funt et al., 2004).

Yellow rust

Yellow rust (*Phragmidium rubi-idaei* (DC) Karst.) is another rust disease that is common around the world in red raspberries. In susceptible cultivars, especially those from Scotland, this disease is effective in causing severe rusting on leaf undersides in autumn and premature defoliation of plants. Symptoms are also present on leaf under-surfaces in springtime and on fruit in a susceptible cultivar, especially under conditions favoring infection. Resistance to this disease is found in many cultivars; it is only a significant issue with very susceptible cultivars like 'Glen Clova'.

Leaf spot

Leaf spot (*Sphaerulina rubi* Demaree and Wilcox), also known as raspberry leaf spot, septoria leaf spot (*Septoria rubi* Westand) and cane spot, attacks susceptible red, black and purple raspberries under warm, rainy weather; it causes leaves to drop and may predispose plants to winter injury. It occurs mainly in North America. Small greenish-black spots develop on the upper surface of the leaf and become whitish or gray, and with time these spots may fall out and create a 'shot hole' effect. The fungus over-winters on canes and affected leaves that have fallen. Spores are produced from these tissues and are distributed by splashing rain to infect young canes and leaves (Bushway et al., 2008).

Powdery mildew

Powdery mildew (*Sphaerotheca macularis* (Wallr: Fr.) Lind.) of red, black and purple raspberries occurs in North America and Europe, but is not found in Australia or New Zealand. It is especially serious when crops are grown under high tunnels and/or greenhouses or in any environment that has hot, dry weather. Powdery mildew infection varies widely between susceptible and resistant cultivars. Infected leaves can be covered with the white powdery growth of the fungus and/or develop green blotches on the upper surface which may be confused with mosaic virus. The leaves may curl or the plant may produce small leaves or become stunted with flower buds and fruit seriously affected (Jennings, 1988). The fungus overwinters on the buds and as shoots emerge from these buds in the spring, the spores are spread by wind and reinfection occurs over the summer. To prevent this disease, avoid planting susceptible cultivars (Table 11.1).

Cane blight

Cane blight (*Leptosphaeria coniothyrium* (Fuckel) Sacc.) is a serious disease of wounds inflicted by machines, hoes, mechanical harvesters, pruning or training wires, or by primocanes rubbing against old (dead) fruiting canes on raspberries in Europe, USA and New Zealand. It is seen on black raspberries as a result of summer heading (tipping) and occurs on red raspberries. It can occur at the same time and in conjunction with spur blight (Table 11.1). Infections begin in the summer as dark areas beneath the wounds on primocanes and can run down the entire length of the cane. Spores from the old (dead or spent) canes are released during rains and dispersed by wind and splashing rain. They then germinate and produce more spores. In a bad year in Scotland, a year after machine harvest, yields can be about 30% of normal due to wounds inflicted by harvesters (Jennings, 1988). Pre- and post-harvest fungicide sprays are effective, as is redesigning the plates on the mechanical harvester.

Anthracnose

Anthracnose (*Elsinoe veneta* (Burkh.) Jenkins) or raspberry cane spot is an extremely serious disease of black and purple raspberries and susceptible red raspberry cultivars in many parts of the world (Jennings, 1988). *E. veneta* attacks succulent young growth rather than mature tissues (Plate 11.8). The canes are most susceptible when they are 10–30 cm (4–12 in) high. Infections are not as severe as the canes get older in mid- to late summer; lesions formed in late summer remain small and shallow. In Scotland, the critical months for *E. veneta* are May, June and July. As these spots enlarge, they become oval in shape, turn gray in the center and develop dark raised borders (Plate 11.8). Diseased tissue extends down into the bark, partially girdling the cane. In the following year, fruit that are produced on severely diseased canes may fail to develop to normal size and may shrivel and dry, particularly in a year of low rainfall. Canes that have been girdled die during the winter, while buds near the cankers either die or cause irregular development of fruiting laterals. The fungus overwinters in the bark within the lesions. Placing the spent canes in the row middle, chopping the canes with a mower, and then spraying the clippings in the dormant season with liquid lime sulfur can reduce the spores from the canes before growth starts (Plates 11.9 and 11.10). The production of spores in the spring coincides with the emerging leaves. Spores are rain splashed, blown or carried by insects to the young, succulent, rapidly growing plants, which are susceptible to infection. Symptoms appear about a week later (Funt *et al.*, 2004). The severity of the infection period is based on the temperature and the numbers of hours the canes are wet after the rain starts. Delayed dormant sprays of liquid lime sulfur (or equivalent) plus additional fungicide sprays, such as Captan (common name) or similar fungicides, having good air movement, and/or producing resistant cultivars in high tunnels or greenhouses can reduce the pressure of the disease (Bushway *et al.*, 2008) (Table 11.1).

Spur blight

Spur blight (*Didymella applanata* (Niessl) Sacc.) is a cane disease that affects red raspberries in North America and Europe, but is rarely found on black raspberries. The leaves are the main infection sites; the fungus grows within the petioles and colonizes in the cortex of the cane around the buds and on the underside of the swollen leaf bases, where it kills the cells and induces a chestnut-brown lesion (Plate 11.9). Symptoms appear in early summer and by mid-summer half of the primocanes may be covered with spur blight lesions. These become silvery gray in the winter and may extend to internodal regions. The buds are so weakened and reduced in size that they fail to develop and are prone to be killed by frost. Spores overwinter and then escape into the air during rainy periods from mid-spring to early summer, starting the cycle again (Bushway *et al.*, 2008). Raspberry cultivars vary widely in resistance to the pathogen, as hairy canes tend to be less affected, especially if they are spine free and/or have a waxy bloom. Several Asiatic species show resistance (Jennings, 1988). Spur blight control is similar to that for control of anthracnose.

Cane Botrytis

Cane Botrytis (*Botrytis cinerea* Pers. F.) or gray mold blight is found in North America and Europe. It can be more damaging than spur blight, can be found in the same locations as spur blight, and mainly affects red raspberries, but can affect other *Rubus* species (Jennings, 1988). Like spur blight, *B. cincera* infects the leaves and petioles in late summer and appears as pale brown leaf spots on mature or senescent primocane leaves. Leaves may shed prematurely. In spring, young lesions are hard to distinguish but as they mature, spur blight becomes a much darker brown while cane Botrytis displays a water-marked appearance and healthy green tissue is found after scraping away the surface of the infection (Bushway *et al.*, 2008). The cane lesions are the principal, initial sources of inoculum for infection of flowers and fruit.

Botrytis fruit rot or blossom blight or gray mold

Botrytis fruit rot (*Botrytis cinerea* Pers. Fr.) is the most serious disease on raspberries worldwide when plants are grown outdoors; there is no problem with this disease when plants are grown under cover. It is the same fungus as cane Botrytis and is found mainly on raspberries in the field during development of the fruit from green to red, in periods of wet or humid weather. It can be found on picked fruit held at high humidity and high temperatures. When gray mold forms on fruit in containers, the fungus restricts sales of fresh fruit in markets that are at some distance from areas of population (Plate 11.10). In Canada, some 64–93% of fruit were found to develop the fungus within 40–60 h after harvest, even though the incidence in the field was less than 1% (Jennings, 1988).

A high incidence of gray mold is associated with long intervals between harvest dates and a high rainfall during the 10-day period of the picking season. A reduction in gray mold can be accomplished by picking sound fruit (Jennings, 1988), picking fruit in the early stage of maturity, picking every day or every other day, and placing fresh picked berries into refrigeration, within 45 min of separation from the cane, under conditions that provide air movement around the fruit (such as containers with vents). Careful handling that resists bruising the fruit is important as well as selecting cultivars with tough skin and a firm texture, which resist bruising. Such cultivars show a degree of resistance to gray mold. Harvesting all ripe fruit, particularly toward the center of the rows from the beginning to the mid-point of harvest, also reduces the number of gray mold fruit from touching and spreading from fruit to fruit. Temperatures of 16–27°C (70–80°F) and moisture on the foliage from dew, rain, fog or irrigation (overhead) create ideal conditions for disease development (Funt *et al.*, 2004). Young blossoms are also very susceptible to infection. Infections then move back into the stems. However, if the weather turns dry after the attack on stems and blossoms, the stems will blacken and the fruit may not be affected.

Gray mold is managed by cultural practices that promote rapid drying of flowers and fruit, such as thinning of canes, removal of spent (dead) canes and tying of canes to the trellis during the dormant season, maintaining narrow rows by dormant and pre-harvest, and the use of drip irrigation rather than overhead irrigation. For vigorous black raspberry cultivars, the spacing of plants (75–90 cm/30–36 in) at planting is important for allowing air movement around plants. Next, the timely and regular harvest of fruit to prevent a build-up of spores and the application of fungicides before rainy periods that occur during blossoming and before harvest can be judicious. A dormant spray of liquid lime sulfur is a very important part of reducing the inoculum before shoots and blossoms appear. Just as important as cultural practices, the removal of field heat by refrigeration within 45 min of picking is vital for reducing post-harvest development of gray mold on berries intended for markets in populated areas that require several hours of transport and hours sitting at the point of sale (Bushway et al., 2008). New fungicide technology allows applications at 10% and 50% stage of harvested berries with no harvest interval necessary to reduce the development of gray mold for up to 7 days.

FRUIT AND PLANT DISORDERS

Winter injury

Winter injury can be a serious plant disorder, caused by abnormal temperatures and weather in the fall or spring, fluctuating warm to deep cold in late winter after plants have achieved their chilling requirements (see Chapter 2 and Glossary 1), or the plant's predisposition to low temperatures caused by insects, disease (root rot), nematodes, wind damage, high soil fertility, and/or high, late nitrogen applications, particularly in mid- to late summer. Winter damage can cause injury to overwintering floricanes, but usually not new primocanes. Winter damage usually does not cause injury to the roots, but *Verticillium*-infected plants will show distinct root symptoms in the spring as warm temperatures cause new growth to occur (Bushway et al., 2008).

Healthy floricane red raspberry cultivars can sustain temperatures of −34°C (−30°F) and black raspberries can sustain −26°C (−15°F). Healthy primocane autumn-bearing red raspberries can sustain similar conditions as floricane-bearing cultivars when canes are pruned just above the ground in the spring. However, if the spent canes are removed in the spring and new canes arise from the roots (below the soil line), these primocanes are the most unlikely to be damaged by cold or show winter injury, unless there is root rot or soils have been saturated for long periods of time and roots have been killed. It is best to plant hardy cultivars, avoid frost pockets by planting on slopes with good air drainage, plant on well-drained soils without a high water table, and avoid over-watering and over-fertilization.

White drupelet disorder

This generally occurs on red raspberries under high temperatures, 32°C (90°F), and high ultraviolet light. Drupelets, either as singles or in groups, are 'bleached', remaining white and not developing the red pigment. These berries are suitable for processing, but are not favored in fresh markets such as farm markets or retail stores. Similar symptoms can be due to tarnished plant bug or stink bug feeding and/or powdery mildew. Shading the plants will reduce this disorder; some cultivars do not have a high percentage of fruit showing the disorder. Thus, the economics of shading may increase costs beyond a small percentage of damage. When temperatures are very high, entire fruit may show symptoms of scald, where the fruit seems to be partially 'cooked' and completely ruined. Therefore, shading under high temperatures may be warranted.

INSECTS

Insects (arthropods) can be beneficial, neither beneficial nor harmful, or harmless. As mentioned previously, certain insects can be vectors of diseases, particularly viruses. In general, growers should protect the beneficial insects, such as pollinators (bees) and predators of other pests, and protect plants from the harmful insects that attack the roots, canes and fruits. Mites are not insects, but are found on the underside of the leaves and can cause economic damage, particularly under hot, dry conditions and under cover. Entomologists around the world have studied insects and their control and provide information on degree days for emergence and thresholds (amount of insects to cause economic loss to fruit or plant damage) as to the best control strategies for predators and/or insecticide use.

Insects that affect the root, crown or cane

The raspberry crown borer (*Pennisetia marginata* Harris) or raspberry root borer is a serious pest of all *Rubus* fruit in North America, but does not occur in Europe. It looks like a yellow jacket (family Vespidae), and is a clear-winged moth with black and yellow bands on the abdomen. It takes 2 years to complete its life cycle (Plate 11.11). Adults appear in mid- to late summer (late July and August in the Midwestern USA). Females can be seen on the foliage where they lay reddish-brown eggs (Plate 11.12) on the underside of the leaf; the eggs take 30–60 days to hatch (Funt *et al.*, 2004). The adults die in about 1 week. After hatching the larvae move to the base of the plant and the next spring enter the crown and roots, girdling the new cane before it goes to the root. In the second winter, it is in the root and by summer, the crown can be damaged. The larvae transform into a pupa and the adult emerges in mid- to late summer. The first symptoms are wilting, dying cane foliage and half-grown fruit (Bushway *et al.*, 2008). Growers should destroy infected canes. All wild bramble plants in and near the planting should also be destroyed. If insecticides are available, a heavy application of insecticide (drenching) in the early spring should kill most larvae and spraying the soil with an insecticide in early fall (mid-October to mid-November in the USA) can kill adults.

The raspberry cane borer (*Oberea bimaculata* Olivier) occurs in eastern parts of North America and Europe, is a slender black beetle, 1.2 cm (½ in) in length, with prominent antennae and usually two black dots on a yellow prothorax. It has a 2-year life cycle; adults appear in early to late summer, feeding on the tender green epidermis of cane tips and leaving brown patches or scars (Plates 11.13 and 11.14) (Jennings, 1988). The female creates two puncture rings around the cane, about 1.2 cm (½ in) apart and about 15 cm (6 in) from the cane tip, and deposits an egg between the rings (Bushway et al., 2008). These grubs (larvae) then feed inside the cane and bore down to the base of the cane by fall and into the crown by the next summer. They feed at the crown during the entire second season and into the second winter.

Control can be achieved by removing and destroying the infected portion of the stem a few inches below the wilted part, immediately after the first site of damage. Damaged canes should be destroyed during dormant pruning (Plates 11.15 and 11.16). Insecticides should be applied to control the adults in a late pre-bloom application.

Strawberry clipper or strawberry bud weevil (*Anthonomus signatus* Say) is a small 0.25-cm (0.1-in) long adult, reddish-brown in color with rows of pits or punctures along its back and two white spots with dark centers. The adult overwinters in fence rows, mulch or wooded areas and emerges as temperatures rise above 16°C (60°F). They move into raspberry fields and feed on immature pollen by puncturing the blossom bud with their snouts (Bushway et al., 2008). Some infected flowers buds are girdled and buds dry out and dangle from the stem (Plate 11.17), eventually falling to the ground. In some areas this pest can cause economic damage and pesticide sprays may need to be applied at the pre-bloom stage (same time as for raspberry fruit worm should control the weevil) (Funt et al., 2004).

The red-necked cane borer (*Agrilus rificollis* Fabricus) or flat-headed cane borer appears in eastern North America, is seldom a serious pest and is a slender metallic black beetle about 0.6 cm (¼ in) long with a reddish or coppery thorax (neck) and short antennae; the larvae are white, legless, 1.9 cm (0.75 in) long and flat-headed (Plates 11.18 and 11.19) (Funt et al., 2004). Galls are formed in symmetrical swellings about 1.2–2.2 cm (1–3 in) across, about 30 cm to 1 m (1–4 ft) above the soil line; canes often break at the swelling and larvae reach full size by fall.

If less than 5% of the canes are affected, infected canes can be pruned out and burned, buried or destroyed (Funt et al., 2004). An insecticide can be applied in late pre-bloom (same as for the raspberry cane borer) if more than 5% of the canes are affected, with another spray at or before petal fall until no more adults are found.

The raspberry cane maggot (*Pegomya rubivora* Coq.) is smaller than a housefly and its larvae (maggots) tunnel down the cane and girdle the cane from the inside, similar to the raspberry cane borer but with this insect there are no girdling marks. This damage may occur each year and the cane should be cut off below the infection when the symptoms appear in the summer; infected canes can also be cut and removed during dormant pruning. No insecticide is recommended. It is found in the USA, Canada (British Columbia) and the UK (particularly England) and is occasionally found on red raspberries (Jennings, 1988).

Tree crickets (Oecanthus sp. Order *Orthoptera*) are small green-white insects with a slender body and dark antennae, which can be longer than its body. Both nymphs and adults can be seen on canes in summer. In late summer, females lay 30–80 eggs in the canes about 0.5 m (1.5 ft) from the tip; this weakens canes and they may be broken (Bushway *et al.*, 2008). Diseases can then enter the injured part of the plant. Damaged (particularly spent) canes should be removed and destroyed after harvest. If damage is severe, an insecticide may be applied in late summer (August and September in the Midwestern USA) or as nymphs appears in the spring (Funt *et al.*, 2004).

The raspberry cane midge (*Resseliella theobaldi* Barnes) is a serious pest in the UK and Europe and plays a role in midge blight. The first generation of adults emerge from the soil in spring (May and June in Scotland) and lays their eggs in splits caused by internal growth stress on primocanes on certain cultivars; larvae are found 2 weeks later. They are translucent at first and turn pink to orange as they mature. The pupae spend 2–3 weeks in cocoons; the second generation appears in mid-summer; a third generation in late summer (August to September) is rare. No recommendations have been given on control (Jennings, 1988).

Insects that affect the leaves and fruit

The larger raspberry aphid (*Amphorophora agathonica*), the smaller raspberry aphid (*Aphis rubicola*) and small (leaf curling) raspberry aphids (*Amphorophora idaei*) are found mainly in North America and Europe. They are major vectors of four major viruses of raspberries (Plate 11.20 and Plate 5.10). These aphids feed on the leaves and cause the leaves to curl downward. After feeding on a virus-affected plant for 15–30 min, the aphid then injects the virus into healthy plants and the virus spreads throughout the plant, resulting in symptoms of mosaic, leaf curl and/or stunting and can yield losses of up to 50% (Funt *et al.*, 2004). Nursery stock can be most vulnerable. The larger raspberry aphid transmits the raspberry mosaic virus complex and the smaller raspberry aphid transmits the raspberry leaf curl virus. The larger raspberry aphid is about 0.3 cm (⅛ in) long and is either yellow-green or pale bluish-green; the smaller raspberry aphid is 0.16 cm ($^1/_{16}$ in) long and pale yellowish-green.

Aphids overwinter as eggs and hatch in the spring into nymphs that proceed through several instars; males are produced in the fall and mate with females (most are wingless) to produce the overwintering eggs (Funt *et al.*, 2004). Aphids feed on tissue near the tip. Aphids are attacked by predators and parasitic insects. Breeders have found that raspberries contain genes that resist aphids, such as *A. idaei*; there are four races of *A. idaei*. The cultivars 'Canby', 'Titan', 'Lloyd George' and 'Royalty' are resistant to aphid feeding. To reduce the risk of aphids, destroy all wild brambles within 200 m (600 ft), purchase essentially virus-free plants, and dig and destroy all plants that show virus symptoms in the field. Aphids are generally reduced by heavy rains. Scout for aphids in the late spring and early summer by looking at the leaves and apply an insecticide if more than two aphids per cane tip are showing (exceeding the threshold) (Funt *et al.*, 2004).

A. idaei are small sedentary aphids that occur in Europe, form dense colonies on fruiting canes and primocanes in spring and early summer, feed on the underside of the leaf causing a curl of the leaf and stunting and twisting of the shoot tips which become sticky with honeydew, and produce a large number of winged forms in early summer that migrate to young canes. Several related species, *A. rubifolii* (Thomas) and *A. spiraecola* (Hatch), also occur on raspberries in North America; *A. ruborum* (Borner) occurs on raspberries in Europe and South America. None of these is known to transmit viruses (Jennings, 1988).

Japanese beetles (*Popillia japonica* Newman), rose chafer (*Macrodactylus subspinosus* Fabricius), and green June beetle (*Cotinis nitida* Linnaeus) are insects (scarab beetles) that feed on the leaves, flower buds, and/or berry fruit. All three of these insects have one generation per year (Plate 11.21). Larvae or grubs develop in pastures, lawns or other types of turf. The Japanese beetles are about 1.25 cm (½ in), copper-colored with metallic green markings and white hairs on the bottom, emerge from pupal chambers in the soil in mid-summer and are more devastating on red raspberries than black raspberries. The rose chafer is light brown, 1.25 cm (½ in) long with long legs, emerges in early spring and feeds most commonly on white flowers and foliage, sometimes destroying flower buds and reducing fruit yields. Sandy soils can promote high populations. They feed on grasses and weeds. Green June beetles are 2.5 cm (1 in) long, metallic green on top and brown on the sides, larvae are soft white grubs with six legs and a curled brown head and are most noticeable in the planting or near a site where manure or compost has been spread. These areas attract the egg-laying females and serve as ideal areas for larvae (Funt *et al.*, 2004). Chemical sprays may be needed at the time that these beetles first appear or at late pre-bloom before the blossoms open (Bushway *et al.*, 2008) or during harvest, where it can be forecast that the insects will destroy approximately 20% of the leaves. Insecticides may need to be applied more than once in a few days. Pre-harvest restrictions must be obeyed.

Another scarab beetle, the New Zealand grass grub (*Costelytra zealandica* (White)), also feeds on the foliage of raspberries. The most serious damage is done by the larvae on the roots, where a severe infestation will eat all the finer roots and the root epidermis of the plant up to ground level, killing the plants. Control methods for grass grub include use of insecticides, including diazinon and lannate. Another effective control measure is through the use of biocontrol agents including *Serratia entomophila*, which prevents the insect larvae from surviving.

Picnic beetle or sap beetles (*Glischrochilus quadrisignatus* Say and *Glischrochilus fasciatus* Olivier) are about 0.5–0.6 cm (⅕–¼ in) long, black with four orange-yellow spots on the back or wing covers. There is one generation per year (Plate 11.22). The larva is white with a brown head and about 0.6–0.9 cm (¼–⅜ in). They will overwinter in many different plant covers and as temperatures reach 16–18°C (60–65°F) in the spring, they feed on fungi, pollen or sap of plants. Adults feed on ripe and over-ripe raspberries and other fruit and/or any other fermenting material (Bushway *et al.*, 2008). One method of control is to place a bucket of over-ripe fruit and allow this to attract the beetles and trap them. Sanitation is the best method of control, with the berries picked frequently or at close harvest intervals so that over-ripe fruit does not occur. Keep ripe berries off the ground and bury culled fruit near packing plants (Funt *et al.*, 2004).

The raspberry sawfly (*Monophadnoides geniculatus* Hartig) larvae are spiny pale-green worms. They are about 1.2–1.9 cm (½–¾ in) in length and chew on the edges of leaves as young larvae; older larvae chew on everything except larger veins, causing a skeletonized appearance. The damage can result in considerable loss of yield. The adult (about 0.6 cm/¼ in) is a black four-winged fly with a yellow band on the abdomen and red markings (Funt *et al.*, 2004). The female lays eggs singly on the top and bottom of the leaf and has one generation per year. It is common in North America, is not common and may occur in low numbers in Midwestern USA, but is not found in Europe (Jennings, 1988). However, a second species (*Priophorus morio*) known as the small raspberry sawfly, which has an appearance similar to *M. geniculatus*, but has two or more generations per year, has been reported to be a problem in greenhouse-grown raspberries in northeastern USA (Bushway *et al.*, 2008). Chemical control is suggested at the early pre-bloom and late pre-bloom stage; this application will also control the fruitworm.

The raspberry fruitworm (*Byturus unicolor*), the eastern USA raspberry fruitworm (*Byturus rubi* Barber), the raspberry beetle (*Byturus tomentosus* Degeer) in Europe, or the western raspberry fruitworm (*Byturus bakeri* Barber) (North America) are small, light brown beetles about 0.3 cm (⅛ in) in length. As an adult, it emerges (Plate 11.23) in early spring (late April in the USA and May in the UK), feeding first on the growing point of the primocane as the leaves begin to open and then on the flower buds and young fruit, keeping the drupelet from developing and condemning any sample for fresh or processing markets (Jennings, 1988). After the larvae hatch, they enter the blossom or young fruit. The larvae, which feed for about 30 days, are fully grown by early summer (July in the USA) (Funt *et al.*, 2004). Early ripening cultivars may be more susceptible to the eastern raspberry fruitworm than late ripening ones. Because the larvae fall to the ground in early summer, autumn-fruiting raspberries often escape injury (Bushway et al., 2008). Cultivation of the ground in late summer can control the larvae, but cultivation that injures roots can cause more disease. Chemical control is mostly used at early pre-bloom as blossom buds appear and late pre-bloom before blossoms open. These sprays should control sawflies as well. Biological control may be available in some locations.

Tarnished plant bug (*Lygus lineolaris* Palisot de Beauvois) adults are about 0.6 cm (¼ in) in length, oval, somewhat flat, greenish- (coppery) brown with reddish-brown markings on the wings and a distinguishing small, yellow-tipped triangle on the back. Nymphs resemble aphids but are more active (Bushway et al., 2008). They are pale green and about 0.025 cm (¹/₁₆ in) when they first hatch. Adults and nymphs have piercing/sucking mouth parts and are present on many plants, such as apple and peach trees and strawberries, until a frost in autumn. After eggs hatch in early spring, control weeds and do not mow forage crops, such as alfalfa, when brambles are flowering, because mowing encourages tarnished plant bugs to move (in the USA). They are found on the crop and weeds and then feed on flowers and developing fruit. To monitor the tarnished plant bug population, sample 50 plants in the morning by tapping one flower or fruit cluster over a tray. Control is advised if there are 0.5 insect per cluster. When that threshold is reached, apply insecticides just before blossoms open and again before the fruit begins to color.

The two-spotted spider mite (*Tetranychus urticae* Koch) is a serious pest in North America and Australia, but is somewhat less of a pest in Europe. It is most commonly found in red raspberries, where it increases in numbers in hot weather (Jennings, 1988). The mites vary in color from pale green to crimson red, have eight legs, and are about 0.05 cm (0.02 in) long. Adult females are oval in shape and marked with two large spots, one on each side; they can produce as many as ten generations per year (Bushway et al., 2008). Under heavy infestation, leaves are marked by white stippling or bronzing after the mites have fed (see Plate 6.9). Severe damage can reduce yield and fruit quality, reducing consumer appeal because they look like brown dust on the fruit (Funt et al., 2004). The number of mites may be reduced under heavy rain or soaking sprays under high pressure. Natural enemies (predators), including predatory mites, lady beetle and lacewings, can be purchased from suppliers in the USA. Miticide sprays are applied as the population increases. Reducing certain types of insecticide spray can reduce the loss of some predators and increase the control of mites. Using a 10× magnifier, monitor the underside of leaves; if there are 10–15 mites per leaf, chemical sprays may be justified. Red spider mites (*Panonychus caglei* Mellott and *P. ulmi* Koch) are found in North America and Europe, respectively, and are frequently found as greenhouse pests but are not generally found in the field (Jennings, 1988). Mite damage in the greenhouse may look like mosaic or other disease (see Plate 6.9).

Potato leafhopper (*Empoasca fabae* Harris) occurs in North America. It is bright green and about 0.3 cm (⅛ in) long. Young nymphs are smaller and light green; adults are identified when they move from side to side. Eggs are hatched within the leaves and stems. Nymphs on the underside of leaves and adults are very mobile and attack over 140 species of plants (Bushway *et al.*, 2008). Potato leafhopper injury can be mistaken for herbicide injury, nutrient deficiency or symptoms of viral infection (see Plate 5.9). Margins of affected leaves develop a light yellow color; new growth can be curled downward and stunted. Generally, damage can be found after the mowing of adjacent alfalfa fields in summer as the insect moves into raspberry plantings.

Yellowjackets (*Paravespula*, *Vespa* and *Vespula* species) feed on ripe and/or injured fruit, particularly when weather conditions are dry. Because of their activity, picking can be difficult, particularly for pick-your-own customers. They generally build their nests underground or in old logs. Their numbers peak in late summer (Plate 11.24). They can be discouraged by sanitation, picking all ripe berries and/or removing over-ripe fruit, and not allowing pickers to bring sugary drinks, lunches or other attractants into the field. Traps can be placed around the planting before harvest.

Yellowjackets cannot be effectively controlled by insecticides. There are several species of this group of wasps in Midwestern USA. Yellowjackets are yellow and black wasps about 1.2 cm (½ in) long. Whitejackets and bald-faced hornets are close relatives, are black and white, and aggressive, nasty stingers (Funt *et al.*, 2004).

WEED CONTROL IN RASPBERRIES

As mentioned several times in this chapter, the control of weeds (reductions in weed pressure) can reduce several pests that use weeds as host plants. The management of weeds is discussed in Chapter 9. It is important to control (or eliminate) perennial weeds for 1–2 years before planting and during the establishment year. Planting cover crops reduces weeds, and increases soil tithe and fertility.

Once plants are set, it is almost impossible to reduce perennial weeds in the row without harm to the plants. Herbicides are labeled in the USA for use in the row and do not harm the plant. Some of these are post-emergence and/or pre-emergent types. However, herbicides cannot control weedy plantings entirely. Hand pulling and hoeing may be necessary. Further, as described above, cultivation close to plant roots may cause an increase in disease problems. Thus, chemical weed control after planting can be advisable for economical weed control and increased productivity. Further, as described in this chapter, the control of certain weeds around and near the planting using a systemic herbicide can reduce many diseases from infected plants that can be harmful to an uninfected raspberry planting. Mulches and/or compost that are weed free may be used the first year but are not suggested in the second or consecutive years, particularly if the soil is heavy (clay) because it could cause more root rot. It is imperative in chemical weed control to apply the correct chemical at the best stage of weed growth and at the correct concentration, and finally, make sure there is adequate cover of the weed or soil area.

SPRAY EQUIPMENT SELECTION, SPRAY COVERAGE AND WORKER SAFETY

The selection of equipment for the application of pesticides is a long-term management decision. The purpose of spray equipment is to deliver nutrients (foliar) and pesticides to the desired target, and ultimately produce a quality and safe product for the consumer. There have been many new developments in sprayer technology from many countries around the world during the latter half of the 20th century (Plate 11.25). Therefore, the manager is faced with a variety of sprayer technology decisions, such as tank size, a pump that has the ability to produce a volume of water, a fan that can disperse the air in the canopy and replace that volume with droplets of spray with the correct size of nozzles to cover the area in a timely manner. Thus, this operation must be capable of completely covering the vegetation in the row width without causing drift onto other crops or fields. A small or large grower, who applies in the field, may have several types of spray equipment, from a small 12-l (3-gal) hand-held knapsack or backpack type to a 200-l (50-gal) three-point hitch power take off (PTO) herbicide sprayer to a large 1200-l (300-gal) trailing (one axle) motor-driven air blast sprayer (with or without a tower). Small backpack sprayers may be used for spot treatments of weeds in established plantings or on newly set plants.

The size and type of pump (piston, centrifugal, diaphragm, or diaphragm/piston) must be capable of delivering sufficient sprays per minute per a set travel speed at a specific pressure to the nozzle and then to the canopy and area covered. Recommendations for an amount of insecticide or fungicide are made per hectare (acre) of plants. Recommendations for herbicides are made as per treated hectare (acre) because only the row area is treated and is only a portion of an area of plants. In many cases one-third of a hectare (acre) is covered with a herbicide, thus an amount per hectare can cover 3 ha of plants.

Tractor size must be matched to the size and type of sprayer. As an example the PTO sprayer may require 25 horsepower (hp) to be effective. The engine size or PTO hp rating of the tractor needs to be 25 hp for the sprayer plus 15 hp or more (total 40 hp) so that it can move itself either on level ground or on slopes. The tractor may have to supply electrical, hydraulic or pneumatic external services to the sprayer as well as having a protective cab, which will need air filtration to protect the driver. Sprayers are capable of covering one, two, three or four sides of the row(s) (Plates 11.26 and

11.27). It is recommended that a grower use a separate sprayer of any size for herbicides and one for insecticides, fungicides and foliar nutrients. Applying pesticides indoors, such as high tunnels and greenhouses, will require different equipment. Regardless of whether spraying indoors or in fields, worker protection is vital for human safety.

Covering the plant and soil or leaf canopy with the proper amount of spray is the first step in effective control of pests and weeds. Always follow label directions for restrictions and concerns for the environment and wildlife. Pesticides can degrade rapidly once in the spray tank and should be applied immediately after mixing. Further, the pH of the water has an effect on the effectiveness of the pesticide. A high pH water can be treated to reduce the pH safely and not harm the spray or the targeted plant. Application timing is equally important. As mentioned previously, the application of a spray during the dormant (no leaf) stage and the pre-bloom or late pre-bloom stage of growth can be very critical to producing a quality fruit for the consumer as well as having effective and efficient control of the pest. Proper pruning, tying and row width that allow good air movement can also allow good coverage by pesticides (see Chapter 10). Pre-emergence herbicides are best applied before weeds emerge in the row; this may be best in late fall or early spring, depending on the type of herbicide.

Safe use of sprayer equipment applies to the worker who needs to be protected from moving parts, such as the PTO shaft, fan or pump, to the tractor operator who needs to be protected from spray drift, and to people and plants that can be affected by spray drift (off target). Keep all PTO shafts in good order and covered with guards, have the tractor driver wear protective clothing or be in a cab that is properly filtered, and ensure that the spray does not become wind-blown or produce large droplets, thereby reducing drift onto off-target people and plants (United States Environmental Protection Agency/United State Department of Agriculture, 1991). In the USA, private or commercial applicators can be fined or placed in jail by state or local regulators if pesticides are improperly handled or applied.

SUMMARY

Raspberries are subjected to many insects and diseases. Control of weeds and nematodes before planting can reduce disease pressure and improve yield and quality of the fruit. Starting with disease-free and essentially virus-free, high-quality plants can be one of the most important decisions a raspberry grower can make. Setting the plants into a well-drained soil that is weed free and maintaining good weed control throughout the life of the planting is vital for early returns on the investment. Annual insect, disease and weed control will be necessary as well as drip irrigation before, during and after harvest in areas deficient in rainfall. The harvesting of ripe fruit at the proper time and maintaining short harvest intervals can be one of the best ways of reducing insects and diseases within the planting. Maintaining good sanitation practices, as mentioned in this chapter, and utilizing refrigeration will be beneficial for providing quality fruit to the customer.

REFERENCES

Bushway, L., Pritts, M. and Handley, D. (eds) (2008) *Raspberry and Blackberry Production Guide* (NRAES-35). Natural Resource, Agriculture, and Engineering Service (NRAES), Ithaca, New York.

Funt, R.C., Ellis, M.A. and Welty, C. (2004) *Midwest Small Fruit Pest Management Handbook*. Bulletin 861. The Ohio State University Extension, Columbus, Ohio.

Jennings, D.L. (1988) *Raspberries and Blackberries: Their Breeding, Diseases, and Growth*. Academic Press, London.

Jones, A.T., McGavin, W.J. and Mayon, M.A. (2000) Comparisons of some properties of two laboratory variants of raspberry bushy dwarf virus (RBDV) with those of three previously characterized RBDV isolates. *European Journal of Plant Pathology* 106, 623–632.

Maloney, K.E., Wilcox, W.F. and Sanford, J.C. (1993) Raised beds and metalaxyl for controlling Phytophthora root rot of raspberry. *HortScience* 28, 1106–1108.

12

CROP PRODUCTION

MICHELE STANTON*

The University of Cincinnati, College of Design, Art, Architecture and Planning, Cincinnati, USA

INTRODUCTION

Raspberries are widely known and grown around the world. In general, raspberries prefer a temperate climate in which they receive winter chilling, and while dormant most cultivars are very tolerant of cold temperatures to −20°C (−4°F) or lower. Outside of their preferred climates, raspberries can be grown in areas of extremely cold winters if insulation is provided for the canes (stems) and roots; this can be achieved by burying the entire plant under soil. They can also be grown as annuals in climates where winters are too warm to provide chilling. Raspberries are considered long day plants (refer to Glossary 1).

In North America, raspberries produce the highest yields in a climate having cool, but not cold, winters, a long growing season where temperatures warm up slowly in spring, and moderate summer temperatures (Plate 12.1). For example 'Titan', a well-known cultivar in the eastern USA, typically prefers daytime highs of 15–20°C (59–68°F) (Fernandez and Pritts, 1994) combined with soil temperatures of 25°C (77°F) (Percival *et al.*, 1996). Primocane-fruiting cultivars grow more quickly as temperatures continue to warm; 22C° (72°F) was found to be optimal for 'Autumn Bliss' (Carew *et al.*, 1999). 'Heritage' flowered and fruited best at 29°C (84°F) (Lockshin and Elfving, 1981). Summers with temperatures not much greater than 25°C (77°F) are usually best. High summer temperatures are a limiting factor in commercial raspberry production, although a few cultivars are adaptable to areas of warmer summers (Stafne *et al.*, 2001; Daubeny, 2002). Black raspberries prefer climates with slightly cooler nights than red raspberries (Plate 12.2) (R.C. Funt, 2011, personal communication). Raspberry plants will grow in areas with a significant amount of shade (Ricard and Messier, 1996), but yield of fruit is maximized when plants are grown in full sun.

*michele.stanton@uc.edu

Raspberries tolerate a wide range of soils, but prefer medium textured soils. They will grow in sandy soils if supplemental irrigation and fertilization are provided, and can handle clay soils as long as drainage is excellent. They prosper in soils whose organic matter content has been augmented before planting. The ideal soil pH is 6.0–6.5, but a range from 5.5 to 7.0 is acceptable (Bushway *et al.*, 2008). Plants grown in soils outside that range may experience mineral nutrient deficiencies. Therefore, the selection of cultivars by commercial growers before planting is important for long-term growth and profitability.

Raspberries need regular precipitation and/or irrigation. Although raspberry plants are usually durable, hardy plants in the wild, commercial raspberries should be grown with supplemental irrigation in place. Apart from temperature, appropriate irrigation is the single most important climactic factor for high berry yield (Prive *et al.*, 1993). If soil drainage and mineral nutrition needs are being met, yields continue to improve with soil moisture content above that which is normally needed for other fruiting plants (Goulart, 1989). Many growing areas receive ample annual rainfall, but virtually none has it consistently from cane emergence through the end of the season. A study in Ontario, Canada, showed that growers overestimate the ideal interval between watering, don't apply enough water, and don't take into account that both precipitation and soil moisture status can vary from place to place in the same field (Bushway *et al.*, 2008).

Good air circulation is important for plant health, but raspberries grown in sites with frequent wind above 2.25 m/s (5 mph) will require the installation of windbreaks (shelter belts). If strong winds occur while the plants are in leaf, even if only infrequently, windbreaks are still essential (see Appendix 1). Good results were obtained in Canada with a 50% permeable artificial windbreak (Prive and Allain, 2000).

Most raspberries must be grown with support (see Chapter 10). Support (trellis) is needed to produce clean fruit, to facilitate both machine and hand harvest, and to ensure maximum light penetration into all areas of the canopy. Yields are highly correlated to leaf light interception (Palmer *et al.*, 1987; Braun *et al.*, 1989), which is facilitated by appropriate support and training (Plate 12.3). Refer to Chapter 10 for in-depth information on support systems.

Wild *Rubus* spp. should be eradicated in as wide a radius as possible around a planting, because these wild relatives can act as a reservoir for pests or diseases (see Chapter 11). This should be completed 1–2 years before the establishment of the planting and be monitored throughout the life of the planting by removing the plant using cultivation (rogueing) or an application of a systemic herbicide.

ASSESSING PLANT HEALTH

Once the planting has been established, plants can be assessed to see whether they are developing normally. Some measurements can be taken to evaluate the health of the planting. First, observe the plant's vegetative growth. Although this varies by cultivar, generalizations can be made. In cold-winter climates, field-grown plants can be expected to produce five to 15 vigorous new canes per year after the first 2 years. A 5-year-old plant, for example, should have at least ten floricanes before pruning. Table 12.1 gives ideal cane density for various production systems and spacings (Hughes and Dale, 2011).

Plantings of primocane-fruiting raspberries should be in rows 45–60 cm (18–24 in) wide; canes that emerge from outside these widths should be rogued. Floricane-fruiting raspberries, especially those grown for machine harvest, should be in narrower rows, 15–25 cm (6–9 in) wide. Canes should be at least 1 cm (0.4 in) in diameter and between 1.2 and 2.0 m (4.0 and 6.6 ft) tall, assuming they were not tipped (headed). In milder climates, such as Oregon in the western USA, floricane-fruiting cultivars should reach 2.3–3.0 m (7–10 ft) in length for both types. Cane growth is initially reflective of a plant's nitrogen status (Hart *et al.*, 2006). Canes should be fairly uniform in color. Well-grown floricane-fruiting cultivars do not exhibit the biennial-bearing tendencies sometimes found in other fruits.

Table 12.1. Production systems, spacing and cane density after pruning.

Production system	No. canes/m^2	Distance (m) between rows	No. canes/ linear m of row fruiting	No. canes/ linear m of row vegetative
Primocane fruiting	10–16	1.5	15–24	–
Primocane fruiting	10–16	2.0	20–32	–
Primocane fruiting	10–16	2.5	25–40	–
Primocane fruiting	10–16	3.0	30–48	–
Floricane-fruiting				
Annual system	5–8	2.0	10–16	10–16
Annual system	5–8	2.5	12–20	12–20
Annual system	5–8	3.0	15–24	15–24
Floricane-fruiting – biennial system				
Non-bearing year	10–16	2.0		20–32
Bearing year	10–16	2.0	20–32	
Non-bearing year	10–16	2.5		25–40
Bearing year	10–16	2.5	25–40	
Non-bearing year	10–16	3.0		30–48
Bearing year	10–16	3.0	30–48	

Leaf color is another indicator of plant health. Some redness of new leaves is normal, although maroon tones can also indicate a temporary phosphorus deficiency that is sometimes caused by insufficient uptake in cool soils (Rehm, 2009). Use of chloride-based fertilizers can also cause red coloration across the whole field. Once the air and soils warm in spring, however, fully expanded leaves should be medium to dark green in color. Some cultivars, such as 'Chilliwack', are notable for their light green foliage; many cultivars with yellow or orange colored fruit also have light green foliage.

GROWTH AND FRUITING OF FLORICANE-FRUITING CULTIVARS

In the central USA, the first flowers of floricane-fruiting cultivars open 6–8 weeks after bud break, with flowering continuing for several weeks. The first fruits ripen 4–5 weeks after pollination. While plants are flowering and fruiting in late spring and early summer, new primocanes (non-fruiting canes) should be emerging as well. The timing of primocane emergence in spring varies according to the cultivar and the environment. In most cultivars, primocanes should have reached a height of 30 cm to 1 m (12–39 in) at flowering and 60 cm to 2 m (24–79 in) at the time of fruiting, unless canes have been burned or hand pruned. Later primocane growth is common in cultivars when winter chill requirements have not been adequately satisfied. Primocanes continue to lengthen until mid- to late summer. In the first year of growth, branching of the main primocane is common, occurring when plants are at least 30–60 cm (12–24 in) tall. It is common to have one to three branches, and up to six branches, in some cultivars and in individual vigorous plants.

Well-grown berries of most cultivars separate readily from the torus when hand picked and fully ripe (see Plate 2.2). They also separate easily or very easily in machine harvestable cultivars (Plate 12.1). Retention of the receptacle does not increase post-harvest fruit quality (Perkins *et al.*, 1994). Cultivars suitable for fresh market production are frequently selected for their ease of removal prior to full ripeness without requiring a two-handed harvest at a light red or even white stage of maturity (Plate 12.2). Drupelets are usually cohesive, but lack of cohesion is frequently evident when cultivars like 'Tulameen' are picked at 'fresh ripe' stage, with some drupelets remaining attached to the receptacle adjacent to the calyx.

Most floricane-fruiting cultivars bred for commercial production have firmer fruit than wild raspberries. Recent advances in breeding have also yielded cultivars whose fruit firmness, skin strength and shelf life are sufficient for shipping and marketing through retail fruit stores and supermarket outlets. Many primocane-fruiting raspberry cultivars possess fruit that is softer and of poorer quality than 'Tulameen', 'Glen Ample' or 'Octavia', the standards for fruit quality in the UK and Europe.

GROWTH AND FRUITING OF PRIMOCANE RASPBERRIES

Primocane-fruiting (autumn-fruiting) red raspberry cultivars such as 'Heritage', 'Autumn Bliss' or 'Autumn Britten', or yellow cultivars such as 'Anne', can bear fruit in the late summer of the first year if they are planted in early spring (April in the USA). Primocane-fruiting types will start to flower and fruit once canes have reached their mature size, usually 210 days after plant emergence. The timing of flowering and fruit is also temperature and cultivar dependent. Older primocane-fruiting cultivars exhibit little or no branching except in the first or second year plantings, but newer cultivars like the Driscoll Strawberry Associates (DSA) cultivar 'Maravilla' continue to branch as long as they are grown. Primocane-fruiting red raspberries in New York yielded from 0.839 MT/ha to 10.106 MT/ha (749 lb/ac to 9205 lb/ac) (Weber et al., 2004). A similar study showed yields of 3.752 MT/ha to 7.317 MT/ha (3351 lb/ac to 6534 lb/ac) in the upper Midwestern USA (Yao and Rosen, 2011). Yellow autumn-bearing raspberries can have a yield nearly the same as red raspberries. In California, yields of primocane-fruiting raspberries can be as much as 30 T/ha (26,790 lb/ac) (H.K. Hall, 2011, personal communication).

EXPECTED BERRY QUALITY

Berries should have a moisture content of at least 80% (Bushway et al., 2008). Soluble solids of ripe fruits range between 5 and 15% (Maloney et al., 1998; H.K. Hall, 2011, personal communication). Individual berry size and total yield varies by cultivar, season and location. A typical raspberry fruit grown in New York weighs 2–3 g (9–14 fruits/oz), while fruits grown in the milder climate of the Pacific Northwest and Canada can be 4–5 g (6–7 fruits/oz), and newer cultivars in other parts of the world may be up to 10 g or larger (Plate 12.1).

Color should be appropriate for the cultivar. Berries are technically non-climacteric (Perkins and Nonnecke, 1992), but will continue to change color from white to red after harvest, and can be picked mature-white for long-distance shipment. Cultivars suitable for fresh market should not go dark red after harvest, but exhibit a non-darkening red color like DSA's cultivars.

When evaluating berry quality, watch for fruits that adhere to the torus, are overly small, crumbly or composed of few drupelets, or are pale in color when fully ripe. These characteristics may indicate problems that should be diagnosed and corrected. Regularly occurring white drupelets in otherwise fully colored, ripened fruit may be symptomatic of damage caused when daytime temperatures exceed 33°C (91°F) accompanied by exposure to strong ultraviolet light during the last phases of ripening (Renquist et al., 1989) or be caused by tarnished plant bugs or stink bugs (Funt et al., 2004). In rare instances, crumbly berries are caused by a somatic mutation (Moore and Robbins, 1990), or may be caused by too low temperatures during flowering and fruit set.

Average fruit yield/hectare varies by type and by cultivar. In Illinois (a cold winter, continental climate area of the central USA), floricane-fruiting red raspberries yield an average of 4500 kg/ha (4000 lb/ac), but black raspberries in New York (also a cold winter, continental climate) yield 2.3–6.8 MT/ha (1–3 T/ac) (Weber, 2006). Machine-harvestable cultivars in the Pacific Northwest often yield 13.5 T/ha (6 T/ac), and newer cultivars may produce as much as 20 MT/ha (9 T/ac) (H.K. Hall, 2011, personal communication).

Plants grown under high tunnels can produce significantly more than their field-grown counterparts. In New York, the difference can be up to tenfold (Heidenreich et al., 2009). There are various reasons for this difference. The fruits are protected from many fungal infections because they escape rains during ripening and harvest; the plants are protected from hail, winter wind damage and bud desiccation and plants can be grown at much greater densities with narrow row spacings, and are tended by people instead of machines. They can experience a longer fall cropping period (autumn-fruiting primocanes) because they can be protected from frost. Spent portions of primocanes can also be removed to promote bud break further down the cane, further enhancing yield (Heidenreich et al., 2009). These higher yields are balanced somewhat by increased costs of production and greater management needs.

MANAGING SOIL WATER

It is worth repeating that most growers apply too little water rather than too much. The purpose of irrigation is to maintain water presence in the soil where it is easily accessible to roots. For maximum productivity, plants should not be allowed to experience drought stress. Remember, too, that water use will vary across the planting with differences in cultivar and environment. Water must be regularly applied in sufficient quantity to replace the amount used by the raspberry plants and by adjacent cover crop plants through transpiration, plus the amount of water lost directly through evaporation. Early in the season, more water will be lost through evaporation from the soil than from transpiration, but total water needs will be low to moderate. As the season progresses, the leaf area of the plants increases, flowering and fruiting occur, ambient temperatures increase, and overall water use is much greater, with transpiration being responsible for a large percentage of that water usage (see Chapter 9).

Plants may not show visible signs of water stress until after damage has occurred, so the successful grower will anticipate a plant's needs and irrigate well before that point. In good soil, raspberries have an effective root depth of 61–91 cm (24–36 in). Growers should try to supply enough water so that soil moisture throughout the root zone stays above 50% of field capacity. In general, the amount of water applied to the root zone is more important to total yield than is the delivery method (Bryla *et al.*, 2007), but drip irrigation may boost yield in that it can be used during harvest whereas sprinklers should not be used. In addition, some types of irrigation are better suited to certain climates. In arid regions, raspberries grown with sprinklers may produce more floricanes than those irrigated with drip (Bryla, 2010). Sprinklers are widely available, can be used when water conservation is not critical, and are usually satisfactory. Sprinklers can also be used to protect the flowers and fruit of primocane-fruiting plants from early frosts. In humid climates, however, water applied to foliage can encourage fungal growth, so drip or furrow irrigation is preferable. If sprinklers are used in humid areas, water should be applied early enough in the morning so that foliage can dry by midday. In areas where furrow or flood type irrigation is the only source of water available, the grower will have to make adjustments so that certain areas of the field do not become too dry or too wet. If water conservation is desirable, drip irrigation is the most efficient method usually available. Weeds are also less numerous in non-irrigated areas (Bushway *et al.*, 2008). Drip irrigation systems can also be used for applying fertilizer and nutrients (fertigation) (see Appendix 2).

It is important that plants receive sufficient water during several critical periods. The first such period is during primocane growth in spring. Stressed plants produce shorter, less vigorous primocanes (Hess *et al.*, 1997). Second, plants need sufficient water from flowering through harvest. Water usage peaks as fruit ripens from green to red (Cameron *et al.*, 1993). Water stress during ripening results in smaller and/or fewer fruits. Drought stress can cause dry, crumbly fruits and may also lessen fruit detachment in machine-harvested fields. Later in the season, insufficient water can cause reduced floral bud development, impinging on autumn primocane yields and/or next year's floricane crop (Hess *et al.*, 1997).

Several things should be kept in mind as the grower attempts to optimize irrigation. First, a grower should be familiar with the texture and moisture-holding capacity of his soils (refer to Chapter 9, Table 9.1). Sandiest soils hold the least amount of water; soils with a large fraction of clay hold the most but also hold it tightly. Soils with higher amounts of organic matter are also more water-retentive than soils with very little organic matter. The faster a soil drains, the more frequently it will need to be irrigated (Bushway *et al.*, 2008).

The grower should be familiar with the plant's daily water usage. A preferred method to estimate the daily water usage (or loss), referred to as evapotranspiration (ET or ETo), is to consult ET tables for the particular location. In regions where growers have access to ET data, that information should also be utilized to schedule irrigation.

If ET rates are not available for the area, Table 12.2 will also give a general estimate. It represents ET rates for well-grown, healthy plants that are grown in well-drained soils with adjacent sod middles during mid-season. Higher ranges should be used where there is little or no cloud cover, or for windy periods. For example, if the area is semi-arid and the mean temperature has been about 22°C (72°F), then the daily loss is 4–7 mm (0.16–0.28 in), indicating a need for a water application of 16–28 mm (0.63–1.10 in) on every fourth day. This also assumes that the soil was irrigated to field capacity on day zero.

Irrigation amount and frequency can vary greatly, even among well-grown plants in suitable soils. Modified soil systems require an adaptation of these principles. For example, some commercial blocks need water as often as three to five times daily when light sandy soils are used as a support medium for semi-hydroponic growing systems (H.K. Hall, 2011, personal communication).

In addition to observing ET rates, it is also important to monitor soil moisture status. This is easily done with tensiometers, evaporation pans or digital meters. In practice, when soil moisture approaches 50%, a grower should calculate the amount of water necessary to bring soil moisture back to 100% of field capacity along the width, depth and length of the row, and apply that amount early in the morning, also allowing extra to compensate for the inefficiencies of the particular irrigation system.

Several tensiometers can be placed at different depths in the same location to verify infiltration rates.

Soil moisture status can also be estimated by hand. To do so, the grower should sample soil from several locations in the planting using a long-bladed

Table 12.2. Estimated evapotranspiration rates under different daily temperatures.

Regions	Mean daily temperature (°C)		
	Cool (~10°C)	Moderate (20°C)	Warm (>30°C)
Tropics and subtropics			
Humid and sub-humid	2–3 mm/day 0.08–0.12 in/day	3–5 mm/day 0.12–0.20 in/day	5–7 mm/day 0.20–0.28 in/day
Arid and semi-arid	2–4 mm/day 0.08–0.16 in/day	4–6 mm/day 0.16–0.24 in/day	6–8 mm/day 0.24–0.31 in/day
Temperate regions			
Humid and sub-humid	1–2 mm/day 0.04–0.08 in/day	2–4 mm/day 0.08–0.16 in/day	4–7 mm/day 0.16–0.28 in/day
Arid and semi-arid	1–3 mm/day 0.04–0.12 in/day	4–7 mm/day 0.16–0.28 in/day	6–9 mm/day 0.24–0.35 in/day

trowel, soil auger, or similar tool to pull up a slice of soil to the depth of the root tips. Take a handful of this soil, squeeze it together firmly, and compare its characteristics to those listed in Table 12.3. The ideal sample will approach 100% of field capacity; here again, plants should be irrigated before moisture deficit approaches 50%. With practice, this method can be very reliable.

The grower can use this information to calculate a water budget for his planting. For example, if the plants are growing in sandy loam soil and the rooting depth is 45 cm (18 in), then the soil would hold about 69 mm (2.7 in) water (45 × 1.5; see Table 9.1) within that depth. When half of that amount (35 mm (1.4 in)) has been lost through evapotranspiration (35 divided by 7 mm/day), irrigation should occur on day five to supply enough water to replace the 35 mm (1.4 in) loss. It may be preferable to water slightly more often during flowering and fruiting; applying 25 mm (1.0 in) every 3–4 days, especially because sandier soils may hold less water than estimated. Also, if using an evaporation pan to calculate evapotranspiration, multiply the amount lost by a factor of 1.2 (Bushway et al., 2008).

In the eastern USA, typical water usage (as in the example above) would indicate daily drip irrigation of 68–102 l (18–27 gal) per day per 30 m (98 ft) of row, with the larger amount for older, more established plantings. An application of 25 mm (1 in) depth of water (any delivery method) requires 25 l/m^2 (6 gal/10 ft^2); 2.5 cm (1 in) water depth requires an application of 2.4 l/930 cm^2 (0.6 gal/ft^2). Water should be applied no faster than it will infiltrate the soil.

Table 12.3. Soil moisture interpretation chart (Miles and Broner, 2006).

Soil moisture status	Moderately coarse texture	Medium texture	Fine and very fine texture
100% (field capacity)	Upon squeezing, no free water appears on soil but wet outline of ball is left on hand.		
100–75%	Forms weak ball, breaks easily when bounced in hand.*	Forms ball, very pliable, slicks readily.*	Easily ribbons out between thumb and forefinger.*
50–75%	Will form ball, but falls apart when bounced in hand.	Forms ball, slicks under pressure.	Forms ball, will ribbon out between thumb and forefinger.*
25–50%	Appears dry, will not form ball with pressure.*	Crumbly, holds together from pressure.*	Somewhat pliable, will ball under pressure.*
0–25%	Dry, loose, flows through fingers.	Powdery, crumbles easily.	Hard, difficult to break into powder.

*Squeeze a handful of soil firmly to make ball test.

CARE THROUGHOUT THE YEAR

This section will discuss the regular care given to a raspberry planting in a chronological manner. Certain tasks must be completed while plants are dormant, and before bud break, at the beginning of the growing season. Other jobs should be completed in an orderly progression as the season unfolds. Although these items are covered elsewhere, it is useful to consider them as they occur during the year.

Before cane emergence

In a cool to cold winter climate, the grower must take care of a number of tasks while plants are still dormant and before growth resumes in the spring. Plants should be pruned, trellises inspected and any repairs made, wayward canes should be brought into place, herbicides should be applied, soil should be tested and plants should be fertilized. In cold winter areas of North America, for example, plants resume growth 5–7 weeks before the average date of last frost. Therefore, these tasks should be completed while the weather is still cool, 8–10 weeks before the expected last frost date in spring. Regardless of climate, the following should be done while plants are dormant:

- First, plants should be pruned and canes managed (see Chapter 10).
- Weed management is the next item that needs attention before bud break. Weeds should be kept to a minimum because they compete with raspberry plants for water, nutrients and light, and can serve as hosts for pests and diseases. Cultivation and hand removal should be done at this time. Current guidelines suggest that cultivation not be deeper than 3–10 cm (1.5–4.0 in) because raspberries have many surface roots. Mulches may provide some measure of weed control, but over-mulching can increase the incidence of root rots (R.C. Funt, 2011, personal communication). Pre-emergent herbicides should be applied before early spring weeds can emerge; contact herbicides should be used on existing weeds (refer to Chapter 9 for specific herbicide recommendations). Even in conventionally managed plantings, however, herbicides should not be the primary method of weed control because escapees can become problematic. Avoid off-label use of herbicides. Roundup (glyphosate) can do considerable damage to raspberry plants when not carefully used (Plate 12.4). It should not be sprayed amongst the plants, even during full dormancy, unless a weed problem makes it necessary to kill or risk killing some plants so that the infestation will not spread throughout the block. Its use should usually be limited to infrequent spot applications or wick and wiper applications, again recognizing that it may kill raspberry plants or sod middles in close proximity. For wick and wiper application, mix 1 l (1 qt) of Roundup with

4 l (4 qt) of water (1:4) to prepare a 20% solution, and exercise great care not to let the product touch anything except the target weeds (Weber, 2004). Very few herbicides are acceptable for organic production; be sure to check your country's guidelines. In the USA, scrupulous attention to pre-plant eradication, management of dense, weed-free sod middles, and clean-cultivated rows are advocated in lieu of herbicides (Kuepper et al., 2003).
- Regarding fertility, a combination of soil testing, leaf tissue analysis and grower observation will provide the best assessment of plant nutritional status (Bushway et al., 2008). Soil samples submitted at this time should give laboratories time to return results to the grower prior to fertilization. New fields should be tested for pertinent mineral levels every year. With established plantings, test no less than every 3 years (Bushway et al., 2008). Nitrogen (N) is the most frequent limiting element, and its levels have been positively correlated to yield more consistently than levels of any other minerals (Prive and Sullivan, 1994). Nitrogen applied very early in the season can be beneficially taken up during the initial vegetative growth flush (Kowalenko et al., 2000), although areas where spring rains are ample may experience leaching and less nitrogen becomes available to the plant (Dean et al., 2000). Four other minerals should also be considered. Berries use lower amounts of phosphorus (P) than many other fruiting crops; excessive amounts may cause zinc (Zn) deficiency. Growers whose soil is naturally high in P or who are using poultry wastes as a soil amendment should regularly monitor P soil levels to avoid excess. Potassium (K) is needed by raspberries in large quantities (Bushway et al., 2008), and is particularly important for plant vigor and berry firmness (R.C. Funt, 2011, personal communication). K is also important for good shelf life and post-harvest quality, but plants are quite sensitive to chlorine (Cl) salts. If KCl is applied, foliage of sensitive cultivars can turn red, as mentioned previously. The grower may also wish to test for boron (B) in higher pH or coarse soils, or if soils in the area are known to be deficient; B is necessary for good flowering and rooting. Calcium (Ca) is important for fruit firmness. Fertigation is the preferred way to apply micronutrients (Bushway et al., 2008) where available. Refer to Chapter 8 for soil fertility guidelines.

In the Midwestern USA (cool to cold winter, humid warm summer climate), current recommendations for field-grown berries under conventional management are to side-dress fertilizers in a split application: half of the recommended amount should be applied in early spring just before or at the beginning of cane emergence, and then the other half 4–6 weeks later as plants start to flower (Bushway et al., 2008). Plants grown under high tunnels are typically fertigated to provide most of the nutrients needed for fruit production.

If fruit is to be organically certified, check domestic and foreign market requirements (GLOBALGAP) for lists of approved fertilizer materials and procedures. In the USA, solid organic fertilizers should be applied as early as possible before the growing season to allow for decomposition. Fresh or partially composted manures must also be applied at least 120 days (17 weeks) prior to harvest if the produce or its irrigation water could contact the manure. For example, if primocane-fruiting berries are being produced for a fall-only crop in the northern hemisphere, and the first harvest is anticipated 1 September or later, the latest application of manure would be 1 May. (In the southern hemisphere, an anticipated initial harvest date of 1 March would indicate that manures be spread no later than 1 November.) Other organic fertilizers or manures that are fully and properly composted (and thus not a risk for pathogen contamination) may be spread at any time. Fresh or partially composted manures should not be applied to floricane-fruiting raspberries or dual-cropped primocane-fruiting raspberries at this time (Kuepper et al., 2003). Further, composts and non-standardized amendments can vary greatly in nutrient content. Growers should assume that composts break down slowly, at varying rates, and yield no more than 50% of their nutrient content during the year of application. Poultry manures are an exception, and may yield up to 90% of their nutrient content during the current year.

Farmers typically apply 4.5–9 MT/ha (2–4 t/ac) of compost, although manures differ in their mineral nutrient content. Poultry waste is more concentrated than cow manure, and is applied at half that rate (Kuepper et al., 2003). Municipal solid wastes may also be suitable sources of compost for raspberry production (Warman, 2009). In heavy soils, growers must be careful to apply a shallow coating; in sandier soils, a layer of compost is desirable for soil moisture conservation. Primary mineral nutrient content of various manures is given in Table 12.4. (For further details on the use of other organic fertilizers or soil amendments refer to Kuepper et al. (2003) and Bushway et al. (2008).)

Table 12.4. Approximate NPK values of various animal manures. (Adapted from: Anon., 1998.)

Animal	% Nitrogen	% Phosphoric acid	% Potash
Dairy cow	0.57	0.23	0.62
Beef steer	0.73	0.48	0.55
Horse	0.70	0.25	0.77
Pig	0.49	0.34	0.47
Sheep/goat	1.44	0.50	1.21
Rabbit	2.40	1.40	0.60
Chicken	1.00	0.80	0.39

Early in the growing season

Early in the growing season the water source should be inspected for quality and pump pressure should be checked; components of the irrigation and moisture monitoring systems should be checked for clogs and damage and be repaired as needed.

A delayed dormant spray of liquid lime sulfur is appropriate at bud break in early spring. This should be applied at green tip to 6.4 mm (¼ in) green stage if it can be done in accordance with local laws and guidelines. In the USA, a spray, mixed at a rate of 5–10% (1 l chemical per 10–20 l water, or 1 gal per 10–20 gal), is commonly used. This kills most overwintering fungi before they can damage new canes, and also serves as a protectant against new infection.

A soil drench of Ridomil (Mefenoxam™ (Metalaxyl-M)) is sometimes used in the USA if Phytophthora root rot has been a concern in previous years. Additional fungicide applications may be required for locally prevalent, problematic diseases.

Another task to start early in the growing season is the weekly scouting and observation of the planting. New growers are advised to keep a regular log of their observations, such as the one found at http://pubs.cas.psu.edu/FreePubs/pdfs/uj242.pdf. Include notes regarding recent rainfall, temperature and wind. Note any dry or wet spots in the planting. Watch for weeds that should be eradicated. Scout for insects, pollinators or other beneficial insects on all areas of the canes, flowers, on the soil, sod middles and at the edge of the field. Note any areas where plants are more or less vigorous, or differ in color or other characteristics. Observe leaves for yellowing, red, maroon or brown colors, spotting, crinkling, or any 'different' colors, shapes or patterns of growth.

Mow sod middles regularly once grasses or broadleaf plants are actively growing. Continue mowing throughout the growing season. Mowing reduces soil moisture loss, increases the density of the grass planting, reduces broadleaf weeds from forming seeds, and thus reduces annual weeds and after harvest allows predators to easily find mice or voles. Further, mowing the grass, particularly in autumn-bearing primocane plantings, will reduce moisture and humidity in the field, reducing gray mold.

In early to mid-spring, some growers practice primocane suppression. Largest yields are obtained by selective floricane thinning and removal of primocanes when they are 15–20 cm (6–8 in) tall (Pritts, 2009). Primocane suppression is thought to improve yield by increasing light penetration and air circulation and decreasing humidity in the lower canopy, thus temporarily reducing competition for nutrients (Damijanovich and Treskic, 2007). This can be done manually; the herbicide dinitrophenol works well, although its use is no longer legal in the USA. In some countries the product 'Hammer'

(AI carfentrazone-ethyl) from Food Machinery Corporation (FMC) is used. Primocanes re-emerge after treatment, and will attain sufficient size to bear the following year. Cultivars vary in their response to this treatment. Refer to Chapter 10 for more information.

Organic growers can also make weekly foliar applications of composted manure teas from mid-spring until onset of flowering (Hargreaves *et al.*, 2008).

Flowering and fruiting

Just before flowering begins on floricane raspberries, apply the first nitrogen application (50% of the total annual nitrogen application), if applying a split application (Funt, 1999). The second of the split application should be approximately 30 days after the first. A split application can be beneficial, especially when there have been heavy rains and/or if the soil is sandy and low in organic matter. Fertigation may be a good option at the second application because the demand for plant growth is now for both the current year's production and new plant growth for floricane types. Split nitrogen applications for autumn-bearing primocanes should be made in mid-spring and then approximately 2 weeks prior to flowering. Again, the second application can be made through the drip irrigation system (see Chapter 9).

While scouting during flowering and fruiting, it is important to become familiar with the many insects in the planting. The majority of these species are either benign or beneficial. Pollinators – the various bees, flies and other insects that are relied upon to distribute pollen throughout the planting – and predators – the carnivorous species that feed upon problematic insects – are numerous. A small number of challenging insects may be visiting the planting. In cool to cold winter climates, many of these insects overwinter in the soil as pupae or adults, and newly emerged adults can now be mating and infesting the plantings. Adults and their new larvae may cause damage by eating flowers, fruits and/or roots, by boring holes in canes, or by acting as vectors for other diseases (see Chapter 11). During scouting, take careful note of canes, unusual swellings, puncture marks, color variation, or unexpected flagging of cane ends, which may indicate the presence of serious pests or disease (refer to Plate 11.18b).

Immature stages of insects are often very easy to control, so timely management is important. Management may involve an insecticidal soil drench or late pre-bloom spray for cane borers or fruit worms, a multi-insect orchard spray to reduce insects before fruit formation, and/or release of predatory mites for high spider mite populations. In areas where Japanese beetles are problematic, a spray of carbaryl or pyrethrins + PBO may be necessary. It is important to spray pesticides during the day when bees and other pollinators are in the fields and to avoid using the most toxic pesticides at least a week before and during the time that pollinators are present, so that beneficial pollinators are not harmed (see Chapter 11).

After pollination, it is advisable to apply a fungicide to prevent gray mold, *Botrytis cinerea*. Botrytis may enter the new fruit if the petals senesce or die, but is not visible until losses are noticed at harvest. Botrytis is especially problematic when weather is cool and humid. Some growers use a foliar spray of fungicide plus micronutrients, as B, every 10–14 days, from early bloom through harvest, for enhanced fruit quality (Strik, 2008) (see Chapter 8).

Once the harvest starts, the use of overhead irrigation is detrimental unless temperatures are very high and should cease if possible; water should be applied without wetting foliage. As harvest progresses, take note of any areas where diseased fruit is more prevalent.

Postharvest

In areas with shorter growing seasons or where winter injury may be a problem, nitrogenous fertilizers should not be applied after harvest. Leaf sampling is performed at that time because mineral nutrient values obtained after harvest are relatively stable. Results should be used to help calculate next season's fertilizer requirements (Hart *et al.*, 2006).

In mild winter climates where fungal diseases are prevalent, the floricanes should be removed to the ground after harvest. Otherwise canes should not be removed until late winter. Continue to scout and monitor plantings. Some insect injury will not be visible until later in the season; other insects have multiple generations that can be controlled in mid- and late summer (see Chapter 11).

Some disease damage will become apparent after completion of the harvest. Certain infections that can occur during the moist, mild days of early spring are not visible until later in the season. Look closely for discolored, irregular or shrunken areas along canes (Heidenrich *et al.*, 2009). The purple and gray sunken lesions of anthracnose, for example, are more obvious at that time (see Chapter 11); those canes should be removed and destroyed. Any areas of the planting where abnormal fruits were problematic need to be revisited to check canes carefully for signs of disease. Various other cane diseases and spur blights may be manifest then. They are more prevalent during humid summer weather when plants have been tipped or pruned or if rubbing or other injury has created an open wound in which fungal spores could grow. All possible problems need to be identified and management practices adopted for any issues that warrant action. It is not beneficial to shred, compost or otherwise leave infested canes on site as they will continue to infect the plantings.

Irrigation is important for the remainder of the growing season, especially in Mediterranean climates where rainfall is sparse in late summer and autumn. Primocane-fruiting cultivars undergo flower bud initiation during June and July in the northern hemisphere, and from November to February in the southern hemisphere. In primocane-fruiting cultivars, an increase in available water at that time results in greater berry size, higher berry number, and overall yield (Prive et al., 1993), while a decrease in available soil moisture will greatly reduce future yield (R.C. Funt, 2011, personal communication).

Late season weed control is important after harvest. Organically managed fields should be hand weeded; however, flame-throwers or approved herbicides may also be used, followed by mulching where appropriate (McDermott, 2011). Standard herbicides for weed control can also be used in the fall (Weber et al., 2004). Systemic herbicides can cause damage or death if applied too close to the plant where drift occurs onto the non-target plant (Plate 12.4).

Irrigation after leaf fall may not be necessary. Although soil moisture should stay above 50% of field capacity, in areas of extremely cold winters where raspberries are borderline hardy, winter survival is enhanced where soil moisture is not overly abundant (Hoppula and Salo, 2006). Floral induction may also be promoted by drier conditions during autumn (Crandall and Chamberlain, 1972).

REFERENCES

Anon. (1998) Fertilizer values of some manures. *Countryside & Small Stock Journal* September–October, p. 75.

Braun, J.K, Garth, J.K.L. and Brun, C.A. (1989) Distribution of foliage and fruit in association with light microclimate in the red raspberry canopy. *Journal of Horticultural Science* 64, 565–572.

Bryla, D. (2010) Effects of Irrigation Method and Level of Water Application on Fruit Size and Yield in Red Raspberry. First Year of Production. *American Society for Horticultural Science Short Talk*. Available at: http://ashs.org/db/horttalks/detail.lasso?id=463.

Bryla, D., Kaufman, D. and Strik, B. (2007) Effect of irrigation method and level of water application on fruit size and yield in red raspberry during the first year of production. *HortScience* 42, 1022.

Bushway, L., Pritts, M. and Handley, D. (eds) (2008) *Raspberry and Blackberry Production Guide* (NRAES-35). Ithaca, Natural Resource, Agriculture, and Engineering Service (NRAES).

Cameron, J.S., Klauer, S.F. and Chen, C. (1993) Developmental and environmental influences on the photosynthetic biology of red raspberry (*Rubus idaeus*). *Acta Horticulturae* 352, 113–120.

Carew, J.G., Mahmood, K., Darby, J., Hadley, P. and Battey, N.H. (1999) The effect of temperature, photosynthetic photon flux density, and photoperiod on the vegetative growth and flowering of 'Autumn Bliss' raspberry. *Journal of the American Society for Horticultural Science* 128, 291–296.

Crandall, P.C. and Chamberlain, J.D. (1972) Effects of water stress, cane size, and growth regulators on floral primordial development in red raspberries. *Journal of the American Society of Horticultural Science* 97, 418–419.

Damijanovich, N. and Treskic, S. (2007) Innovation boosts raspberry yields. *Appropriate Technology* 34, 65–67.

Daubeny, H. (2002) Raspberry breeding in the 21st century. *Acta Horticulturae* 585, 69–72.

Dean, D.M., Zebarth, B.J., Kowalenko, C.G., Paul, J.W. and Chipperfield, K. (2000) Poultry manure effects on soil nitrogen processes and nitrogen accumulation in red raspberry. *Canadian Journal of Plant Science* 80, 849–860.

Fernandez, G.E. and Pritts, M.P. (1994) Growth, carbon acquisition and source-sink relationships in 'Titan' red raspberry. *Journal of the American Society of Horticultural Science* 119, 1163–1168.

Funt, R.C. (1999) Improving black raspberry fruit size and yields. *Ohio Fruit ICM Newsletter* 3(15).

Funt, R.C., Ellis, M.A. and Welty, C. (2004) *Midwest Small Fruit Pest Management Handbook*. Bulletin 861. The Ohio State University Extension, Columbus, Ohio.

Goulart, B.L. (1989) Influence of three soil water levels on micropropagated greenhouse grown red raspberry. *Acta Horticulturae* 262, 277–283.

Hargreaves, J., Sina Adl, M., Warman, P.R. and Vasantha Rupasinghe, H.P. (2008) The effects of organic amendments on mineral element uptake and fruit quality of raspberries. *Plant Soil* 308, 213–226.

Hart, J., Strik, B.C. and Rempel, H. (2006) *Nutrient Management Guide: Caneberries*. Corvalis, Oregon State University Extension. Available at: http://extension.oregonstate.edu/catalog/pdf/em/em8903-e.pdf.

Heidenreich, C., Pritts, M.P., Kelly, M.J. and Demchak, K. (2009) High Tunnel Raspberries and Blackberries. Department of Horticulture Publication No. 47. Ithaca, Cornell University. Available at: http://www.fruit.cornell.edu/berry/production/pdfs/hightunnelsrasp2012.pdf.

Hess, M., Strik, B.C., Smesrud, J. and Selker, J. (1997) Western Oregon Caneberry Irrigation Guide. Oregon State University Department of Biosource Engineering. Available at: http://bee.oregonstate.edu/Documents/IrrigationGuides/CANEBERR.pdf.

Hoppula, K.I. and Salo, T.J. (2006) Effect of irrigation and fertilization methods on red raspberry winter survival. *Acta Agriculturae Scandinavica, Section B* 56, 60–64.

Hughes, B. and Dale, A. (2011) The Ideal Raspberry Cane Density. *The Ontario Berry Grower*. Ontario Ministry of Agriculture, Food and Health. Available at: http://www.omafra.gov.on.ca/english/crops/hort/news/allontario/ao0111a2.htm.

Kowalenko, C.G., Keng, J.C.W. and Freeman, J.C. (2000) Comparison of nitrogen application via a drip irrigation system with surface banding of granular fertilizer on red raspberry. *Canadian Journal of Plant Science* 80, 363–371.

Kuepper, G.L., Born, H. and Bachmann, J. (2003) *Organic Culture of Bramble Fruits Horticulture Production Guide*. NCAT Sustainable Agriculture Project. Available at: https://attra.ncat.org/attra-pub/summaries/summary.php?pub=15.

Lockshin, L.S. and Elfving, D.C. (1981) Flowering response of 'Heritage' red raspberry to temperature and nitrogen. *HortScience* 16, 527–528.

Maloney, K.E., Reich, J.E. and Stanford, J.C. (1998) 'Encore' red raspberry. *New York's Food and Life Sciences Bulletin* 152. Available at: http://ip.cctec.cornell.edu/techdocs/D2255.Encore_Raspberry.pdf.

McDermott, L. (2011) Late Summer Weed Control Options for Berry Crops. *New York Berry News* 10(7).

Miles, D. and Broner, I. (2006) *Estimating Soil Moisture*. Colorado State University Extension Factsheet 4.700. Colorado State University, Fort Collins, Colorado. Available at: http://www.ext.colostate.edu/pubs/crops/04700.html.

Moore, P.P. and Robbins, J.A. (1990) Maternal and paternal influences on crumbly fruit of 'Centennial' red raspberry. *HortScience* 25(11), 1427–1429.

Palmer, J.W., Jackson, J.E. and Ferree, D.C. (1987) Light inception and distribution in horizontal and vertical canopies of red raspberries. *Acta Horticulturae* 173, 159–166.

Percival, D.C, Proctor, J.T.A. and Tsuita, M.J. (1996) Whole-plant net CO_2 exchange of raspberry as influenced by air and root-zone temperatures, CO_2 concentration, irradiation, and humidity. *Journal of Horticultural Science* 121, 838–845.

Perkins, P., Strik, B.C. and Collins, J.K. (1994) Retention of raspberry receptacles does not increase fruit quality. *HortScience* 29, 463.

Perkins, V.-P. and Nonnecke, G. (1992) Physiological changes during ripening of raspberry fruit. *HortScience* 27, 331–333.

Pritts, M.P. (2009) Pruning raspberries and blackberries. *New York Berry News* 8(4).

Prive, J.P. and Allain, N. (2000) Wind reduces growth and yield but not net leaf photosynthesis of primocane-fruiting red raspberries (*Rubus idaeus* L.) in establishment years. *Canadian Journal of Plant Science* 80, 841–847.

Prive, J.P. and Sullivan, J.A. (1994) Leaf tissue analyses of three primocane-fruiting red raspberries (*Rubus idaeus* L.) grown in six environments. *Journal of Small Fruit and Viticulture* 2, 41.

Prive, J.P., Sullivan, J.A., Proctor, T.A. and Allen, O.B. (1993) Climate influences vegetative and reproductive components of primocane-fruiting red raspberry cultivars. *Journal of the American Society for Horticultural Science* 188, 393–399.

Rehm, G. (2009) *Soil Temperature*. University of Minnesota Extension, Minneapolis/St Paul, Minnesota. Available at: minnseotafarmguide/agbuzz.com.

Renquist, R.A., Hughes, H.G. and Rogoyski, M.K. (1989) Combined high temperature and ultraviolet radiation injury of red raspberry fruit. *HortScience* 24, 597–599.

Ricard, J.P. and Messier, C. (1996) Abundance, growth and allometry of red raspberry (*Rubus idaeus* L.) along a natural light gradient in a northern hardwood forest. *Forest Ecology and Management* 81, 153–160.

Stafne, E., Clark, J.R. and Rom, C.R. (2001) Leaf gas exchange response of 'Arapaho' blackberry and six red raspberry cultivars to moderate and high temperatures. *HortScience* 36, 880–883.

Strik, B.C. (2008) *Production Practices for Maintaining High Yield and Quality of Raspberries for Processing*. PowerPoint Presentation given at Chilealimentos. Available at: http://www.chilealimentos.com/medios/Servicios/Seminarios/2008/Berries/Strik-Frmabuesa.pdf.

Warman, P.R. (2009) Soil and plant response to applications of municipal solid waste compost and fertilizer to Willamette raspberries. *International Journal of Fruit Science* 9, 35–45.

Weber, C.A. (2004) Raspberry weed management. *New York Berry News* 3(4).

Weber, C.A. (2006) *Black Raspberry Potential, Pitfalls, and Progress*. Cornell University, Ithaca, New York. Available at: http://www.hort.cornell.edu/grower/nybga/reports/Weber%20Black%20Raspberry%20Expo%20Proceedings.pdf.

Weber, C.A., Maloney, K.E. and Sanford, J.C. (2004) Long-term field performance of primocane fruiting raspberry cultivars in New York. *HortTechnology* 14, 590–593.

Yao, S. and Rosen, C.J. (2011) Primocane-fruiting raspberry production in high tunnels in a cold region of the Upper Midwest United States. *HortTechnology* 21, 429–434.

13

POSTHARVEST PHYSIOLOGY AND STORAGE OF RASPBERRIES

PENELOPE PERKINS-VEAZIE*

Plants for Human Health Institute, Department of Horticultural Science, North Carolina State University, North Carolina, USA

INTRODUCTION

Raspberries are delicate fruits and are highly perishable for many reasons. They are made up of many individual fruits (drupelets) held together by hairs (trichomes) and waxes (Mackenzie, 1979). Raspberries lack a protective cuticle or rind, so they lose weight rapidly when moisture moves from the fruit into the surrounding drier air. Unlike blackberries, which detach complete with their central receptacle in place, raspberries detach from the receptacle, leaving a cavity in the center of the fruit. This cavity fruit leads to a lack of stability when raspberries are placed on top of each other in containers and so they are easily crushed. Ethylene, a ripening hormone that often causes softening, is produced in the receptacle and to a lesser degree in the fruit, and this also contributes to perishability. The respiration rate of raspberries is high and contributes to moisture and sugar loss. Lastly, ripe raspberries are highly susceptible to gray mold (*Botrytis cinerea*). This fungus infects flower stamens and then appears inside the fruit following harvest (Plate 13.1). Gray mold growth is hastened at storage temperatures above 5°C (41°F), especially if the fruit has been grown outdoors and has been subjected to dew or rainfall. Gray mold is much less of a problem if the fruit has been grown in a tunnel or a greenhouse. Color change, from pink to red to dark red, can be very rapid if air temperatures are warm (>30°C, 86°F), with a pink fruit becoming dark red in less than a day. Dark red fruit may not be attractive in some markets. All of these fruit attributes make postharvest handling and storage of raspberries a real challenge.

Steps in the postharvest life of raspberries start long before the final product is ready for market/consumption. One of the first decisions to make is market selection. Fruits destined for commercial or long-distance markets

*Penelope_perkins@ncsu.edu

are often different from those used for direct or local markets. Commercial marketing will require considerable thought and attention to refrigeration, storage, cold chain and marketing, and use of cultivars with longer shelf life. In contrast, direct marketing requires attention to timing of availability of fruit, large berry size, high productivity, ease of harvesting and good flavor.

SELECTION OF VARIETIES FOR TARGETED MARKETS

For long-distance fresh markets, raspberries have to remain firm, have light and bright color when fully red, and appear fresh rather than shriveled. Raspberries grown for fresh market have to be able to retain firmness during harvest, handling and storage and have a shelf life of 10 days or more (Toivenon et al., 1999). Genotypes differ in inherent firmness; firmness at harvest can be increased by picking raspberries at the white or pink stage, but picking at these stages requires a cultivar that is easy to pick early, does not crumble, and develops flavor early. Genotype firmness appears to be higher with heavier berries that have more drupelets (Robbins and Moore, 1991). Keeping fruit firm and light in color is easier when harvested under cool, low humidity conditions. After harvest, berries should not be left in direct light, and when conditions are warm, it is essential to frequently take harvested fruit and place into refrigerated storage (cold room or cool store) for field heat to be removed. Fresh market raspberries are harvested at a less ripe stage than fruit designated for processing. The most outstanding cultivars for fresh market in the USA and around the world are those from Driscoll's Strawberry Associates (DSA), Plant Sciences, Naturipe and Five Aces. In Europe, these cultivars are also used either directly or under license; there are also a number of other breeding programs associated with commercial producers developing high-quality cultivars for their proprietary use. In the public domain, some of the better fresh market varieties are 'Glen Ample', 'Octavia', 'Tulameen', 'Tadmor', 'Korpiko', 'Nova', 'Killarney' and 'Joan J'. 'Himbo Top', which is a light colored, glossy fruit, does well in the cool climates of the northern USA but becomes very leaky when harvested in the warmer southern climates (P. Perkins-Veazie, 2010, unpublished data). 'Joan J' turns a very dark, almost purple-red when fully ripe, but remains firm for 7 days after harvest if held below 5°C (41°F). The new cultivars 'Imara', 'Kwanza' and 'Kweli' (available through Meiosis) offer a significant advance for growers of high-quality, fresh market fruit without the need to commit to growing proprietary varieties.

Local or regional markets include direct (farmer's market or roadside stand), pick-your-own (PYO) (customer comes to the farm to harvest fruit), or community supported agriculture (CSA), where customers pay before the season for a specific amount of fruit delivered to them. These customers often prefer a berry high in flavor, but a less firm berry can be used because of a shorter interval from harvest to consumer.

As well as the usual productivity issues, such as yield and disease resistance, variety selection depends on production climates. Raspberry fruit size can decrease as air temperature increases (Remberg et al., 2010), with loss in fruit size also dependent on variety. Selection of raspberry varieties for direct marketing can depend greatly on local availability and also customer familiarity with raspberries. For instance, customers used to plentiful high-quality raspberries may want something more unique, such as a yellow raspberry or extra-large berry size. Raspberries grown for processing need to have a degree of fruit firmness, ease of harvest, excellent flavor and aroma, dark color, and low drip loss (juice loss with thawing).

Raspberries produce ethylene as they ripen. The ethylene produced appears to help abscission of the fruit from the torus (receptacle), and is concentrated in this area (Burdon and Sexton, 1990a; Perkins-Veazie and Nonnecke, 1992). Raspberry cultivars differ in the amount of force required to detach berries from receptacles (Hall et al., 2002), which may be due to the relative amount of ethylene produced in the receptacle (Iannetta et al., 2000). Ethylene can also be produced in fruit tissue, with amount dependent on genotype (Burdon and Sexton, 1990b). Raspberries differ greatly in storage life and relative postharvest problems primarily due to genetic differences, followed by production environment and postharvest handling. The amount of ethylene produced appears to be correlated to higher respiration and decreased postharvest quality (Robbins and Patterson, 1989; Robbins and Fellman, 1993).

HARVESTING

Fresh market

The strength of raspberry fruit adhesion to receptacle tissue decreases as fruits change color and ripen (Sjulin and Robbins, 1987). Berries designated for fresh market are picked at a pink to light red stage, when fruit can be removed from receptacles by hand without tearing the berry. For fresh or direct type markets, raspberries are gently tugged from the receptacle and placed directly into packages in the field. Unlike tree fruit, fresh market raspberries are handled only once, from the plant into the final container, and are not washed.

Because raspberries are soft, have high respiration, and high ethylene production, harvesting for fresh markets is best done when air temperatures are cool (Burdon and Sexton, 1994). For commercial operations in the USA, raspberries are picked quickly into plastic clamshells (Plate 13.2), placed in master containers (cardboard cartons with holes in the sides for forced air cooling and reinforced corners for stacking), then taken to a quality control area to check for US Department of Agriculture (USDA) standards compliance. USDA standards include few or no unripe fruit, no plant debris such as stems, leaves or calyxes, no mold or rust, no visible insects, and a generally uniformly colored and sized pack. Once passed, cartons are taken to cooling facilities, either into refrigerated trucks in place in the field, or to nearby refrigerated storage. Some growers have found that all-terrain vehicles or golf carts can be used for quick trips to the cooler with small loads.

Processing

Sometimes fresh market growers will harvest overripe berries or soft fruit for processing as a means of cleaning plants and recovering part of the production costs. In the USA, most of the processing acreage is harvested by machines. Berries designated for processing are picked at a full to dark red stage, when fruit adhesion is minimal and machines can easily dislodge berries by using a shaking system. Fruit falls on catch plates, moves via conveyor belts across an air cleaner and hand-sorting belts, then drops into containers. The rate of speed of the harvester needs to be slow enough to avoid damage to canes and to fruit; too fast a speed will mean more stems and green fruit.

Training for hand harvesting

If harvesting by hand for fresh, PYO or other markets, people have to be trained to pick at the desired stage of ripeness. Long-distance shippers may want only pink to light red, while PYO growers may want people to pick from red to dark red stages. Picking an area clean of ripe fruit is critical in both types of operations in order to decrease disease and maximize returns. All pickers need to be shown how to remove fruit without breaking off vegetative shoots or leaving receptacles in the berries. Because raspberries are fragile, care must be taken not to grasp fruit too hard, to avoid bruising or leakage. Also, injured or discolored berries should never be picked into packs designated for shipping; moldy berries should be harvested and discarded. Berries are usually picked into cups or clamshells in the field, and should be placed gently in the container, not dropped. Only a few layers of raspberries can be put into a container to avoid crushing the bottom fruit, and pickers must also avoid overfilling clamshells, as snapping the hinged lid on overfull boxes will smash the fruit.

Sanitation

Good agricultural practices (GAPs) enter into postharvest decisions for all types of markets, as well as in decisions for production areas. Raspberry fruit must be clean when picked as washing of fresh market fruit will not be done.

Animal entrance into fields and packing houses needs to be none to minimal, and canes may need to be trimmed or workers trained to avoid picking fruit close to the soil line. Irrigation must be with potable water, and preferably applied in a way to avoid wetting berries. Sanitary areas, including washing facilities, need to be available for customers or professional help and within easy access of the plots being harvested. If drive areas in fields are dusty, then materials brought in or out of the field should be kept covered. Packing areas, box storage and cooling facilities should be kept clean. Boxes should be covered with tarpaulins or held in closed areas to avoid dust, birds and insect contamination. Refrigerated rooms should be swept and washed down as needed to minimize debris or mildew growth. Packing sheds need to be swept to avoid dust.

A more comprehensive list of guidelines and procedures to maximize food safety can be found in Kader (2002) and links to GAPS, HAACP can be found at http://www.gaps.cornell.edu/weblinks.html. A summary list of guidelines adapted from Kader (2002) is given below:

1. Be careful using animal or biosolids waste to avoid microbial contamination. Refer to NOP 5006 (USDA National Organic Program, 2011) for correct compost temperatures and application times.
2. Follow all applicable laws, regulations and guidelines for agricultural practices.
3. Make sure all sources of water that might come into contact with raspberries at any point are safe, including irrigation.
4. Avoidance of microbial contamination is much more effective than trying to remediate contamination.
5. Raspberries can become contaminated anywhere in the food chain from field to table. The major sources of microbial contamination are associated with human or animal feces.
6. Worker hygiene and sanitation practices in all aspects of production, harvest, storage, handling, transport and marketing is vital to minimize microbial contamination.

Packaging

Over the years, packages for raspberries have consisted of buckets, plastic, wood or pulp (composite) boxes, and more recently clamshells. Raspberry US box size is generally kept at 114, 170 or 228 gal (4, 6 or 8 oz) size. Occasionally, a large (342 gal, 12 oz) shallow box is used for the restaurant trade. Usually raspberries are placed in two to three layers to minimize crushing from berry weight. In the USA, plastic clamshells without sharp internal edges or corners, especially around ventilation holes, are the most preferred for best shelf life (Plate 13.2). The clamshells are plastic boxes with an attached lid, with small air holes on the bottom and top of the box. Clamshells have an advantage because of the extra rigid structure in the corners that allow several boxes to be stacked without crushing fruit (Plate 13.3). Clamshells are designed to be placed into cardboard containers (master containers) made to hold 12 or 24 clamshells (total of 1.5–3.1 kg (3.3–6.9 lb) per container) that have additional rigidity and special vents for forced air or room cooling (Plate 13.4).

All containers should be recycled or disposed of, rather than being reused for raspberries. Mold spores readily enter the cardboard of masters and composite boxes. GAPs for sanitation should be followed to prevent transfer of soil, dirt or microorganisms.

Processing fruit are packed into plastic lugs or crates that can hold 10–20 kg (22–44 lb) fruit per container. Growers using fruit for farmers' markets or CSAs may harvest into US ½-pint (~114 g) pulp baskets, bamboo baskets or foam baskets. Wood or composite boxes are sometimes used in direct marketing to accentuate fresh appearance. These containers have several problems for raspberries. They tend to remove water from fruit, which may be beneficial if fruit are leaky, but also leave stains on the package. Ventilation for efficient cooling is limited, especially in wood splint boxes, and the sharp edges of wood splint boxes can easily cut the berries caught in the corners. A wrap is usually placed over the top of containers to prevent weight loss and held in place with a rubber band, requiring hand labor.

Cooling and storage

Rapid cooling followed by constant refrigerated storage remains the most effective way to prolong raspberry shelf-life. Raspberries can be held for up to 14 days under ideal conditions. These conditions include selection of a firm cultivar, proper harvest, handling, cooling and storage protocols, and maintenance of the cool chain through transit and distribution. The best temperature for raspberries is –0.5°C (31°F), actually slightly below freezing. The high sugar and acid content of raspberries acts as an antifreeze, keeping berries from solidifying. Respiration and ethylene production are greatly reduced at 0–5°C (31–41°F) (Table 13.1). Berry color remains red longer at a temperature closer to 0°C (32°F) (Nunes, 2008), and gray mold spore germination and mycelia growth are slowed at temperatures below 5°C (41°F). As storage temperature increases, shelf-life is greatly shortened. Worse, if pre-cooling is delayed, shelf-life is even shorter, regardless of subsequent storage temperature. In fruit harvested at 20°C (68°F), a delay of as little as 2 h in time to access pre-cooling to 5°C (41°F) cut shelf-life by 4 days (Moore and Robbins, 1992).

Raspberries have the best shelf-life if cooled to 5°C (41°F) within an hour of harvest, and if kept between –0.5°C and 0°C (32°F) and 95%

Table 13.1. Estimated raspberry shelf-life, assuming 90–95% relative humidity. Adapted from Perkins-Veazie (2002).

Storage temperature (°C)	Days shelf-life	(CO_2) Respiration rate (mg/kg/h)
−0.5	12	na[a]
0	10	na
1	8	na
2	6	16–18
3	5	na
4–5	4	18–27
10	2	31–39
15	1.5	28–55
20	1	74–175

[a]Not available.

relative humidity during storage (cool store). This means that in almost all production areas, some sort of cooling system and/or refrigerated storage will be needed.

Cooling of produce depends on the density in the container, the vent and types of containers, the volume to surface area, the travel distance of cooling air (shorter distance means faster cooling), and airflow capacity (more airflow means faster cooling) (OMAFRA, 1998). Raspberries packed in clamshells and vented containers mean low density, uniform cooling surface, high surface area to volume (raspberries are small relative to apples), and therefore, will take less cooling time than the same weight of apples.

Room cooling is generally done in cold rooms, and is effective if small amounts of fruit are harvested and the room is large enough to have good air circulation. Containers with well-placed vents should be placed in rows with at least 30 cm (12 in) between rows, and box fans placed to direct cold air through the boxes and tunnels, creating a serpentine cooling system. Often efficient cooling is decreased by frequent door opening, when cold air can quickly be replaced by warm air. Plastic strips placed over the doorway help block loss of air, as well as entrance of insects, birds or other animals.

Cooling coils of the refrigeration system in a cold room have to be cooler than the air. If the temperature difference is large then heat transfer rate is greater and smaller cooling coils can be used (Boyette et al., 1991). But, as coils cool, more water vapor from the air will condense on the coils, as ice or moisture. This is a big concern where air coming into the room is hot and humid and refrigeration must cool air and condense water vapor; sometimes the amount of ice from water vapor on the coils can stop refrigeration. Also, with raspberries, higher humidity is needed. In this case, the best system would be large evaporator coils and small temperature drops across the coils (OMAFRA, 1998), by using a forced air cooler inside a refrigerated room.

Cold rooms can be purchased from commercial contractors, built by growers, or recycled from other industries (Thompson and Spinoglio, 1996). New commercially available cold rooms are expensive; costs can be saved by buying used rooms or by grower construction. Prefabricated cold rooms are less costly, often used in restaurant industries, and can be assembled fairly easily, but must be enclosed inside another structure. The marine containers used on ships make good cold rooms as they are well insulated and generally have refrigeration controls added. If constructed by growers, proper insulation is critical to maintain effective temperatures, with moisture resistant (Styrofoam or polyurethane) plus spray-in foam giving additional insulation (Boyette et al., 1991; http://storeitcold.com/). All cold rooms require effective refrigeration units with the size of unit dependent on room size and heat load (Btu or J/h).

Room cooling is not effective at rapidly removing field heat (the amount of heat in fruit when it comes in from the field), especially in warm temperatures and with many loads coming into the room over a day. Cold air from evaporator coils goes over and around the raspberries and slowly removes heat (Thompson et al., 2002). Minimum airflow needs to be 100 cfm (cubic feet per minute)/ton or 0.3 m^{-3}/ton of product storage capacity (Thompson et al., 2002). Because raspberries do not tolerate moisture (wetting), the most effective means of removing field heat is by room cooling or to use forced air cooling followed by storage in a cool (refrigerated) room. Cool room temperatures for raspberries should be between 1 and 4°C (34–40°F). When raspberries have been harvested under warm conditions (greater than 15°C (59°F)) rapid removal of field heat is critical to avoid fungal growth and color change (see Chapter 11). For these berries, a forced air system is necessary. The forced air is best accomplished by a small room (cool room) inside a larger cold (refrigerated) room. Further, such a cool room can be set up in a corner (cold wall) of a large refrigerated room. This system is more efficient for larger loads.

Forced air cooling should be done inside a cold room. Once raspberries are cooled, they should be moved to another cold (refrigerated) room (cool store) to maximize space for the forced air system. Forced air cooling can be done by building a cold wall or by building home-made systems (Thompson and Spinoglio, 1996; OMAFRA, 1998). A very simple system is to place a box fan at one end of a tunnel made of two rows of containers or pallets, so that the box fan is pulling cold air from near the room fans, place a tarpaulin over the cartons at the other end, and push warm air out of the cartons into the room, utilizing the negative pressure created by the tunnel. Exhaust air from the fans should be moved away from room cooler coils, and the pallets nearest the fan housing should be pressed into foam strips to create an air seal. Forced air cooling can quickly dry out fruit, especially at the top of the pallet, so a layer of cardboard is often placed over these top cartons to block airflow over the top of the fruit (Plate 13.5).

With large loads of raspberries, a cold wall is more efficient. Here, a wall is built about 1 m (3 ft) from the cooler wall and set points are cut in the wall to place pallets of fruit. A maximum fan rate of 1–2 cfm/lb or 0.001–0.002 m^3/sec.kg is advised; moving air volumes above this speed increases the static pressure and energy consumption without greatly affecting cooling rates (Thompson *et al.*, 2002). Forced air cooling can quickly dry out fruit, especially at the top of the pallet, so a layer of cardboard is often placed over these top cartons to block airflow over the top of the fruit (Plate 13.6).

Raspberries have to be held under high humidity in order to reduce weight loss and avoid shriveling, yet have to be held below 100% relative humidity to avoid mold growth. Containers or covers that cut down on the surface area of exposed raspberries help hold in humidity once fruit are fully cooled. Effective methods can be as simple as a clean plastic tarpaulin over pallets or boxes. Fine mist systems can be built into cold rooms to aid humidity. In some older rooms where wooden boards were used for floor or wall construction, wet, clean coarse cloth, such as burlap, can be placed on the floors. Also, controlling airflow during cooling and keeping cooling times within 3 h helps reduce weight loss (OMAFRA, 1998).

For fruit designated for freezing, heat removal and storage are done in the same way as explained above. After cooling, the fruit is washed and sorted to eliminate stems, leaves, insects, unripe bruised or moldy berries and soil. Fruit may go either to solid pack of 8–10 l (8–10 qt) containers or pureed and then frozen or placed on a belt through a freezing tunnel for rapid freezing as IQF (individual quick frozen) to maintain the shape of individual berries and then kept in a large room freezer (cold store) and held at −20°C to −23°C (−5°F to −10°F) (see Chapter 14 and Glossary 1).

CALCULATING FIELD HEAT, AMOUNT OF COOLING TIME AND REFRIGERATION NEEDED

The specific heat of a fruit depends on composition and is in units of heat per weight per unit temperature. Raspberries are 85% water, so specific heat is usually rounded up to 1, where 1 Btu of heat must be removed to cool 1 lb of raspberries by −17°C (1°F). To calculate field heat in a load of raspberries, the following formula is used:

Field heat (Btu/h) = SH [fruit specific heat (Btu/lb/°F)] × DT [difference in field temperature (°F) − desired temperature (°F)] × W [weight (lb)]

SH: 1.0
DT: 50 (pulp temperature of raspberry is 82°F and desired temperature is 32°F)
W: 381 lb (weight of load is 381 lb (171 kg)) for one pallet of 90 containers, each containing 12 half-pints (160 g)

then amount of field heat to remove is 1 × 50 × 381 or 19,050 Btu/h

if want to cool in 12 h then 19,050/12 = 1588 Btu needs to be removed each hour.

The amount of refrigeration capacity (RC) = Btu/h from above/ refrigeration per hour.

One ton of refrigeration is 288,000 Btu/24 h or 12,000 Btu/h. So, the amount of refrigeration capacity will be 19,050 Btu/h/12,000 Btu/h ton or 1.59 ton refrigeration units running constantly.

If more fruit is added to the cooler (refrigerated facility or cool store) then the additional heat load must be calculated, plus the respiration rate of both uncooled fruit and cooled fruit must be added in (Boyette *et al.*, 1991). To reduce the amount of refrigeration needed in a room cooler, one can instead partially cool fruit with forced air then place the fruit in the cold room.

For forced air cooling of raspberries, the recommended airflow rate is 4 cfm/lb or 4 l/s/kg for 1 h cooling to reach ⅞ of the final temperature, or 1.25 cfm/lb for 2.5 h to reach ⅞ of the final temperature (OMAFRA, 1998).

COOL CHAIN

The ideal way to move raspberries for markets is to have pre-cooled refrigerated trucks dock directly against loading docks, with doors opening into a cooled landing area with direct access to cool rooms. Pallets of cold fruit are moved by forklift into the trucks without being exposed to warmer air. Pallets need to be stacked correctly to allow for good air circulation from the top and sides of the truck. Refrigeration should be held at 2–4°C (36–39°F) or lower throughout the journey, which may last 3–7 days. Trucks should be unloaded at distribution points in the same fashion, opening directly into cold rooms or refrigerated reception areas. To ensure that nothing goes wrong with this system, larger producers often own and operate their own refrigerated trucks and delivery vehicles for delivery to the point of sale.

Trucks need to be adequately pre-cooled before being loaded and shipment temperatures need to be monitored through to the receiving agent. Refrigeration in trucks needs to be well serviced and refrigerant loading kept within design specifications; any refrigerant leaks need to be repaired without delay. Loads must be well secured, and cooling needs to be adequate to accommodate delays at truck checkpoints or in dense traffic. Temperatures in trucks must not be allowed to rise upon arrival. If temperatures range from 5 to 8°C (41–46°F) in the freight area upon arrival, depending on placement, then care should be taken to rapidly sell fruit received at higher temperatures as quality will have been compromised.

Shipping mixed commodity loads is common in some parts of the USA, especially in areas where only a few palettes of raspberries are needed at a distribution point. In these cases, fruit or vegetables needing warmer temperatures, such as tomato or watermelon, should not be shipped with raspberries. Breaking the cool chain can reduce expected shelf-life of raspberries by 2 or more days, and in extreme cases can cause load rejection. Additionally, ethylene-generating fruit, vegetables or flowers should not be in mixed loads with raspberries, or ethylene scrubbers should be added to packs or to the truck. Vibration injury that causes fruit to move up and down or side to side will quickly cause berry leakage. Trucks should have balanced tires and minimal trailer vibration. Smooth roads, without potholes and sudden dips, to the highway help decrease transit injury, as does lower speed on gravel, dirt or poorly kept roads.

EXTENSION OF SHELF-LIFE

Germplasm incorporating slow or minimal (non-darkening) color change and significant fruit firmness is an effective means of increasing shelf-life. New selections should be tested for their shelf-life and suitability for transportation as some components of these traits in raspberries are innate and cannot be directly recognized. Simple methods to reduce heat load include using light colored packaging, placing raspberries in shaded areas or providing shade, and making frequent trips to coolers. Rapid cooling, refrigeration, maintenance of cold chain and suitable packaging are the most commonly used and successful postharvest methods for extending shelf-life (Banados *et al.*, 2002).

Use of ethylene or ethylene-promoting substances and conditions should be avoided as it can promote undesirable color change, softening and mold growth. Ethylene scrubbers can be added to storage units as air filters or sachets. These are usually potassium permanganate pellets that react with ethylene gas and can be monitored for loss of activity by color change from purple to brown. Activated charcoal may also be used, and it will scrub carbon dioxide as well as ethylene. Large storage rooms (over 15,000 ft^3) may benefit from photocatalytic units that use ultraviolet light to break down ethylene.

Despite the high respiration rate and rate of decay in raspberries, modified atmosphere storage must be used with care (Agar and Strief, 1996). Raspberries held under 10–15% CO_2 to slow gray mold development and color change can have objectionable flavor, depending on the cultivar (Toivenon *et al.*, 1999). Volatiles indicative of fermentation (acetaldehyde, ethyl acetate, ethanol) increased in 'Meeker' and 'Qualicum' after 7 days' storage in 10% CO_2 and 5% O_2 and enhanced loss of firmness in 'Qualicum'. When held at 6% CO_2 and 10% O_2, fermentative volatiles were greatly reduced (Toivenon *et al.*, 1999). Harvesting raspberries with an intact receptacle and pedicel helps retain fruit firmness, as the cortex provides substantial structural support. The stems can present an injury hazard to the fruit, however.

SUMMARY

In spite of all this attention and care through harvest and handling, consumers end up being a major cause of raspberry quality loss. Raspberries may be purchased in ideal shape then held in a hot vehicle for an hour or two, or placed on a kitchen counter overnight. Mold spores often set within the cavity of the raspberry and wait for ideal conditions to germinate. Consumers may find their beautiful box of raspberries covered with gray mold within a day. Consumers should be made aware of such conditions and maintain quality by proper handling, such as placing cool berries in a paper bag for the trip to their home.

The key for successful postharvest life in fresh market raspberries lies in decisions made long before harvest. Proper cultivar selection for the desired market needs to be determined, with large, light colored, firm fruit essential for long shipping distances. Production environment is very important in determining labor and cooling needs for fresh markets. Where day temperatures exceed 30°C (86°F) for 6 h or more, raspberries can ripen within a day. Fresh market berries must be harvested gently by hand into the appropriate container, given rapid cooling to slow softening and color change, and held at cold temperatures through the shipping and marketing steps.

REFERENCES

Agar, I.T. and Strief, J. (1996) Effect of high CO_2 and controlled atmosphere (CA) storage on the fruit quality of raspberry. *Gartenbauwissenschaft* 61, 261–267.

Banados, M.P., Zoffoli, A.S. and Gonzalez, J. (2002) Fruit firmness and fruit retention strength in raspberry cultivars in Chile. *Acta Horticulturae* 585, 489–493.

Boyette, M.D., Wilson, L.G. and Estes, E.A. (1991) Design of room cooling facilities: structural and energy requirements. *AG-414-2, North Carolina State University.* Available at: http://www.bae.ncsu.edu/programs/extension/publicat/postharv/ag-414-2/index.html.

Burdon, J.N. and Sexton, R. (1990a) The role of ethylene in the shedding of red raspberry fruit. *Annals of Botany* 66, 111–120.

Burdon, J.N. and Sexton, R. (1990b) Fruit abscission and ethylene production of red raspberry cultivars. *Scientia Horticulturae* 43, 95–102.

Burdon, J.N. and Sexton, R. (1994) Practical implications of differences in the ethylene production of *Rubus* fruits. *Acta Horticulturae* 368, 884–892.

Hall, H.K., Stephens, M.J., Alspach, P.A. and Stanley, C.J. (2002) Traits of importance for machine harvest of raspberries. *Acta Horticulturae* 585, 607–610.

Iannetta, P.P.M., Wyman, M., Neelam, A., Jones, C., Taylor, M.A., Davies, H.V. and Sexton, R. (2000) A causal role for ethylene and endo-b-1,4-glucanase in the abscission of red-raspberry (*Rubus idaeus*) drupelets. *Physiol Planta* 110, 535–543.

Kader, A.A. (2002) Postharvest biology and technology: an overview. In: Kader, A.A. (ed.) *Postharvest Technology of Horticultural Crops*, 3rd edn. University of California, Davis, California, pp. 39–47.

Mackenzie, K.A.D. (1979) The structure of the fruit of the red raspberry (*Rubus idaeus* L.) in relation to abscission. *Annals of Botany* 43, 355–362.

Moore, P. and Robbins, J. (1992) Fruit quality of stored fresh red raspberries after a delay in precooling. *HortTechnology* 2, 468–470.

Nunes, M.C.N. (2008) *Quality of Fruits and Vegetables*. Blackwell Publishing, New York, pp. 167–179.

OMAFRA (Ontario Ministry of Agriculture, Food and Rural Affairs) (1998) *Tunnel Forced-Air Coolers for Fresh Fruits & Vegetables*. Ontario Ministry of Agriculture, Food and Rural Affairs, Guelph, Canada. Available at: http://www.omafra.gov.on.ca/english/engineer/facts/98-031.htm.

Perkins-Veazie, P.M. (2002) Raspberries. *USDA Handbook 66*. Available at: http://www.ba.ars.usda.gov/hb66/contents.html.

Perkins-Veazie, P.M. and Nonnecke, G. (1992) Physiological changes during ripening of raspberry fruit. *HortScience* 27, 331–333.

Remberg, S.F., Sonsteby, A., Aaby, K. and Heide, O.M. (2010) Influence of postflowering temperature on fruit size and chemical composition of Glen Ample raspberry (*Rubus idaeus* L.). *Journal of Agriculture and Food Chemistry* 58, 9120–9128.

Robbins, J. and Fellman, J.K. (1993) Postharvest physiology, storage and handling of red raspberry. *Postharvest News and Information* 4, 53N–59N.

Robbins, J. and Moore, P. (1991) Fruit morphology and fruit strength in a seedling population of red raspberry. *HortScience* 26, 294–295.

Robbins, J. and Patterson, M. (1989) Post-harvest storage characteristics and respiration rates in five cultivars of red raspberry. *HortScience* 24, 980–982.

Sjulin, T.M. and Robbins, J. (1987) Effects of maturity, harvest date and storage time on postharvest quality of red raspberry fruit. *Journal of the American Society for Horticultural Science* 112, 481–487.

Thompson, J. and Spinoglio, M. (1996) Small-scale cold rooms for perishable commodities. Publication 21449, University of California, Davis, California. Available at: http://ucce.ucdavis.edu/files/datastore/234-701.pdf.

Thompson, J.F., Mitchell, F.G. and Kasmire, R.F. (2002) Cooling horticultural commodities. In: Kader, A.A. (ed.) *Postharvest Technology of Horticultural Crops*, 3rd edn. University of California, Davis, California, pp. 97–112.

Toivonen, P., Kempler, C., Escobar, S. and Emond, J.-P. (1999) Response of three raspberry cultivars to different modified atmosphere conditions. *Acta Horticulturae* 505, 33–38.

USDA National Organic Program (2011) Guidance: processed animal manures in organic crop production, NOP 5066 pp.1–3. Available at: http://www.ams.usda.gov/AMSv1.0/getfile?dDocName=STELPRDC5087120.

10.1

10.2

10.3

Plate 10.1. Red raspberries being harvested by a pick-your-own customer. (Courtesy of The Ohio State University.)
Plate 10.2. Removal or heading of a new (primocane) black raspberry tip with a gloved hand. (Courtesy of The Ohio State University.)
Plate 10.3. Black raspberry (cv. 'Jewel') in early spring after spent canes have been removed and laterals are thinned and cut back. (Courtesy of R.C. Funt, Carobeth Berry Farm, Ohio.)

Plate 10.4. Raspberry harvester in a trellis support system. (Courtesy of The Ohio State University.)
Plate 10.5. Raspberry harvesters can pass over wooden trellis posts without damage to the harvester. (Courtesy of The Ohio State University.)

Plate 11.1. Verticillium wilt symptoms on black raspberry plant. (Courtesy of The Ohio State University.)
Plate 11.2. Bluish streaks on black raspberry canes affected by Verticillium wilt. (Courtesy of The Ohio State University.)
Plate 11.3. Above-ground symptoms of Phytophthora root rot on primocane of 'Heritage' red raspberry. (Courtesy of The Ohio State University.)
Plate 11.4. Below-ground symptoms of Phytophthora root rot and crown rot on red raspberry. Note the red line between the top healthy green tissue and the infected reddish-brown tissue of the root. The reddish-brown discoloration of the root is typical of root rot. (Courtesy of The Ohio State University.)

Plate 11.5. Crown gall on the root of an infected red raspberry plant. (Courtesy of The Ohio State University.)
Plate 11.6. Orange rust on the underside of black raspberry leaf. (Courtesy of The Ohio State University.)
Plate 11.7. Plants showing small, yellow leaves on healthy plant infected by orange rust. (Courtesy of The Ohio State University.)
Plate 11.8. Anthracnose lesions (cane spot) on black raspberry canes. (Courtesy of The Ohio State University.)

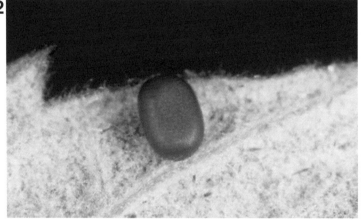

Plate 11.9. Symptoms of spur blight on red raspberry canes. (Courtesy of The Ohio State University.)
Plate 11.10. Botrytis fruit rot, gray mold on red raspberry. (Courtesy of the Ohio State University.)
Plate 11.11. Adult raspberry crown borer. (Courtesy of The Ohio State University.)
Plate 11.12. Raspberry crown borer egg. (Courtesy of The Ohio State University.)

Plate 11.13. Raspberry cane borer damage. (Courtesy of The Ohio State University.)
Plate 11.14. Raspberry cane borer damage. (Courtesy of The Ohio State University.)
Plate 11.15. Dormant prunings (laterals and spent canes) of black raspberry in the drive row ready for mowing. (Courtesy of R.C. Funt, Carobeth Berry Farm, Ohio.)
Plate 11.16. Dormant black raspberry prunings chipped with rotary mower, which is also used to mow the sod drive row between black raspberry rows. (Courtesy of R.C. Funt, Carobeth Berry Farm, Ohio.)

11.17

11.18

11.19

11.20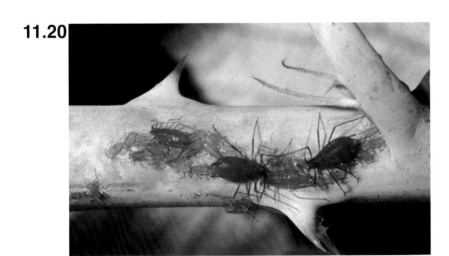

Plate 11.17. Strawberry clipper damage on raspberry. (Courtesy of The Ohio State University.)
Plate 11.18. Red-necked cane borer adult. (Courtesy of The Ohio State University.)
Plate 11.19. Red-necked cane borer damage on cane. (Courtesy of The Ohio State University.)
Plate 11.20. Raspberry aphids. (Courtesy of The Ohio State University.)

Plate 11.21. Japanese beetle on leaf. (Courtesy of The Ohio State University.)
Plate 11.22. Adult picnic beetle. (Courtesy of The Ohio State University.)
Plate 11.23. Raspberry fruitworm adult. (Courtesy of The Ohio State University.)
Plate 11.24. Yellowjacket wasp. (Courtesy of The Ohio State University.)

11.25

11.26

11.27

Plate 11.25. Three-point hitch sprayer on a narrow tractor designed for spraying raspberries in narrow rows; it can apply sprays to two sides in one pass. (Courtesy of The Ohio State University.)
Plate 11.26. Sprayer applying herbicides to young raspberry plants, to two sides of the row plus drive row. (Courtesy of Enfield Farms, Washington State.)
Plate 11.27. A three-point hitch PTO air blast sprayer applying a dormant spray to three sides of a black raspberry plantation near dusk. (Courtesy of R.C. Funt, Carobeth Berry Farm, Ohio.)

12.1
12.2

12.3

12.4

Plate 12.1. 'Cascade Delight', from the Washington State University breeding program at Puyallup, Washington, is a high-quality, large red raspberry for hand and mechanical harvesting. (Courtesy of David Karp, University of California, Riverside.)
Plate 12.2. A black raspberry fruit cluster at harvest. The red colored berry is immature and does not pull easily from the receptacle and is unsuitable for market. (Courtesy of The Ohio State University.)
Plate 12.3. Red raspberries supported by plastic string. (Courtesy of Harvey K. Hall.)
Plate 12.4. Roundup (glyphosate) herbicide damage to raspberry. (Courtesy of Harvey K. Hall.)

13.1

13.2

13.3

Plate 13.1. Moldy raspberries create a bad flavor and mold can spread to other berries. (Courtesy of R.C. Funt.)
Plate 13.2. Clamshell boxes stacked in a grocery store display. Notice slits on top of boxes for added ventilation and cooling, and slightly depressed top of clamshell to allow for stable stacking. (Courtesy of R.C. Funt.)
Plate 13.3. Clamshell showing shallow-size box. Raspberry fruit is equal to the height of the clamshell. (Courtesy of R.C. Funt.)

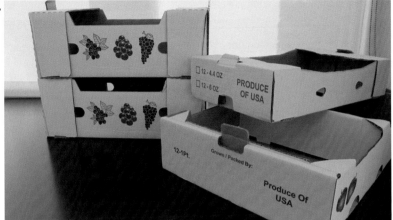

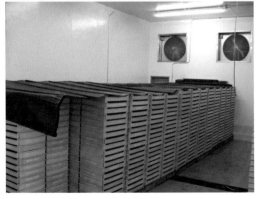

Plate 13.4. Cardboard containers (often called 'masters' in the USA) showing reinforced corners, side vents for cooling and interlocking tabs to help keep containers stabilized in a pallet stack. (Courtesy of Penelope Perkins-Veazie.)

Plate 13.5. Pallets of raspberries placed for forced air cooling using a false wall plenum for pushing out air. (Courtesy of Penelope Perkins-Veazie.)

Plate 13.6. Pallets of berries for fresh market showing side vents in cardboard master containers and room cooling with fans and placement of pallets. (Courtesy of Penelope Perkins-Veazie.)

14.1a

14.1b

14.2

14.3

Plate 14.1. Shiny, firm and damage-free red **(a)** and black **(b)** raspberry fruit. (Courtesy of The Ohio State University.)
Plate 14.2. Cardboard master with six empty quart clamshells. (Courtesy of R.C. Funt.)
Plate 14.3. From left to right: quart, pint and half-pint clamshells showing hinged lids. (Courtesy of R.C. Funt.)

14.4
14.5

14.6

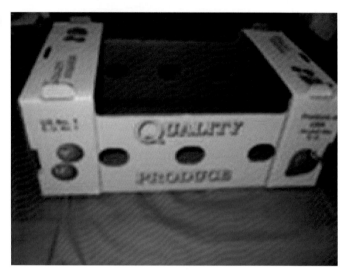

14.7

Plate 14.4. Cardboard master with 12 empty pint clamshells. (Courtesy of R.C. Funt.)
Plate 14.5. Cardboard master before folding. (Courtesy of R.C. Funt.)
Plate 14.6. Cardboard master ready for clamshells. (Courtesy of R.C. Funt.)
Plate 14.7. Containers (lugs) on mechanical harvester in Washington State. (Courtesy of Enfield Farms, Washington State.)

15.1

15.2

15.3

15.4

Plate 15.1. Mechanical harvester in Washington State. (Courtesy of Enfield Farms, Washington State.)
Plate 15.2. Mechanical raspberry harvester in New Zealand. (Courtesy of Harvey K. Hall.)
Plate 15.3. 'Autumn Britten' cultivar is a large berry suitable for high tunnel production. (Courtesy of C. Weber, Cornell University.)
Plate 15.4. Sprayer applying herbicides to young raspberry plants. (Courtesy of Enfield Farms, Washington State.)

Plate 16.1. Raspberry processing plant in China. (Courtesy of Harvey K. Hall.)
Plate 16.2. Fresh picked berries in China. (Courtesy of Harvey K. Hall.)
Plate 16.3. *Rubus crataegifolius* harvested from the wild for commercial sale in China. (Courtesy of Harvey K. Hall.)
Plate 16.4. High tunnel raspberry production in Chile. (Courtesy of Harvey K. Hall.)
Plate A1.1. The trees on the left are willows (*Salix matsudana*). Windbreak (shelterbelt) trees at the other end of the rows are Lombardy Poplar. Windbreak species on the right include eucalyptus species. Row spacings in the raspberries are 3 m (10 ft) and post heights are ~2 m (6 ft). (Courtesy of Harvey K. Hall.)

MARKETING OF RASPBERRIES

GAIL NONNECKE,[1*] MICHAEL DUFFY[2] AND RICHARD C. FUNT[3]

[1]Department of Horticulture, Iowa State University, USA;
[2]Department of Economics, Iowa State University, USA; [3]Department of Horticulture and Crop Science, The Ohio State University, USA

INTRODUCTION

The major components of a business are production, management and marketing (selling) of items for a profit. In today's terms, the value chain of raspberries from production through sales includes the critical step of marketing. The major components of marketing are placing the product into a bulk or retail container, labeling and delivering to local, retail or wholesale markets. For raspberry production to be profitable, selling of quality fruit at fair prices creates a sustainable enterprise, and markets provide the opportunities for sales. Recommendations indicate the importance of marketing in advance by stating that raspberries should be sold before production is started or when fruit is first harvested. Determining and developing appropriate markets are critical for sales of highly perishable raspberries. It is vital to a successful raspberry operation to determine how the crop will be marketed.

The diverse market channels for raspberries provide unique opportunities for distribution and sales. In the USA raspberries are produced for the fresh or processing markets. Fresh markets include on-farm sales that are either pick-your-own (PYO) or hand-picked berries for community farmers' markets, community-supported agriculture (CSA) or pre-picked (hand-picked) berries for wholesale (large stores) or roadside markets. Processed markets include berries that are immediately frozen or made into jams or jelly and juices. A majority of processed berries in the USA are mechanically harvested. However, many PYO berries are processed in the home immediately after harvest, into pies or jams, and/or frozen for consumption at special events or holidays. Also, home gardens produce berries that are either eaten fresh or processed for later use, as gifts or for sale at specialty markets. Raspberries can

*nonnecke@iastate.edu

be red, black, purple or yellow. This diversity allows growers to offer different types for niche markets and to offer many 'value-added' products. Variables that are controlled by the grower include the value of the fruit (price), quality, form and packaging associated with the fruit and its sales (product), publicity to sell the product (promotion and advertisement) and location of the distribution and sales (place).

To determine price, growers need to consider the costs associated with producing their raspberries and the amount needed to obtain a profit (Funt, 1990) (see Chapter 15 for costs of producing raspberries). Growers must keep accurate records of variable and fixed production costs to be used in determining appropriate prices for sale of the raspberries. Additional factors include knowing the amount customers are willing to pay and any competitor's prices. Selling raspberries below the costs of production to undercut competitors does not contribute to a profitable and sustainable operation and should be avoided (Bushway *et al.*, 2008). A grower's cost of producing and marketing red raspberries in the future can be estimated. The most important costs to consider are labor and fixed costs and/or overhead costs. Consideration must be given to making a return on your management time, hired labor and costs of capital. Also, the cost per unit (pint, crate, master) is dependent on how many units are actually sold. One general estimate for the field is that 80% of the expected yield will be sold in either direct market or wholesale markets and that 20% could be lost to birds, mold or spoilage. Raspberries produced under protective covers may have a marketable yield of 92–95%.

FRUIT QUALITY AND MARKETING EXPERIENCE

A high-quality product contributes to successful marketing of raspberries, and the statement that 'quality sells' is as accurate for raspberry fruit and its products as for any consumer product. Quality of fresh raspberries includes fruit that receives top grade, such as US No. 1, and is based on freedom from mold, decay, sun scald, over-ripeness and injury (Perkins-Veazie, n.d.; USDA, 1992). High-quality raspberries are evenly colored, firm and hold their shape with all drupelets intact. Fruit that are shiny indicate freshness and will draw repeat customers (Plates 14.1a and 14.1b).

Factors of quality associated with marketing and sales of raspberries include some aspects that are separate from the fruit itself. If a grower sells fruit through agricultural tourism experiences, the experience must also be of high quality. Attractive, clean and welcoming farm markets and PYO farm conditions can showcase raspberries and raspberry products and provide a pleasant and high-quality entertainment or tourist experience.

Promotion of raspberries is facilitated through a relationship with the public and customers. Its goal is to encourage demand for and sales of raspberries and raspberry products. Promotion strategies are unique to each grower and seller and include any aspects related to public relations, advertising through mass media, direct mailings, the internet and social media and sponsorship of grower or non-grower events. Determining who will buy the raspberry fruit allows for targeted promotion. Customers can range from those visiting a farm for a PYO experience to large-scale processors purchasing large volumes of raspberry fruit.

The raspberry market is the location of the fruit's distribution. For customers to purchase raspberries, the fruit must be available at the right place at the right time. Efficient and effective locations are important for growers to meet their goals for sales. In the planning process for direct marketing, growers should determine how many units of land are already in the market, the population within the area and the impact your supply of berries will have on the market (Courter and How, 1990).

Markets that are used for raspberries and raspberry products include selling directly to the consumer or through wholesale channels. Direct-to-consumer markets have the potential to provide good returns to the grower because the grower is the distributor of the fruit and products and no other intermediary is involved. Direct markets include selling pre-picked fruit at community farmers' markets, on-farm markets, off-farm markets and through markets of CSA enterprises. PYO operations sell fruit directly to the consumer, but the customers harvest the raspberries at the grower's farm.

Fresh pre-picked raspberries offered directly to the consumer are sold in smaller quantities, typically in one-half or one-pint containers that are made of biodegradable pulp, vented plastic (hinged containers called 'clamshells') (Plate 14.2), meshed plastic or wood. Shallow or smaller containers are better because the weight of the top fruits can damage the quality of the lower fruits. Wooden containers are usually placed into 24–36 quart crates and clamshells are placed into 6-quart or 12-pint masters (Plates 14.3 and 14.4). In the USA, masters (Plates 14.5 and 14.6) are made of cardboard, which are easily folded together for clamshells, easily reused or discarded and recycled.

DIRECT-TO-CONSUMER MARKETS FOR RASPBERRIES

Community farmers' markets bring customers to a central location for the convenience of the shoppers, such as in an urban setting of a town or city, where farmers offer products for sale (Sabota et al., 1980). Most farmers' markets are organized whereby growers benefit from having customers come to one location (marketplace), and the sponsoring location also benefits from its support of the farmers' market by increasing customers' activities in the town or city. Major considerations for a farmers' market include location, facilities, parking, operational procedures, including days and hours of operation, stall fees, publicity and legal requirements, such as permits, items to be sold and the methods of sales (Sabota et al., 1980). Many local farmers' markets are part of associations that facilitate the development and organization of community farmers' markets and training of market managers. Commercial growers often participate in several farmers' markets across multiple communities so that they can sell raspberry fruit daily or at least several times in a week, because fresh raspberries decline in quality after storage of several days.

On-farm markets are retail operations located on the grower's farm that offer raspberries and raspberry products for sale. Highly perishable raspberries benefit from on-farm market sales where they can be stored in refrigeration (see Chapter 13 on postharvest management). Markets should include an attractive sales area that is clean and staffed by competent and welcoming workers, a car parking area, proximity to major roadways and convenient hours of operation to increase on-farm market sales (Bushway et al., 2008).

Off-farm markets are operated by the grower but located at a site different from the farm. The off-farm market may provide a better location for customers than the farm site, such as being located along major roadways or intersections. These markets also require an attractive sales area with sufficient space for parking and convenient hours. The off-farm market structures may be permanent facilities, typical of an on-farm market, or temporary. If off-farm facilities are temporary, such as a sales shed, refrigerated storage still should be available because raspberries are highly perishable.

In the USA, CSA is a partnership between the grower and the community, who work together to establish a local food system (Gradwell et al., 1999). In a CSA, community members agree to a full season price before production occurs, sharing with the grower the risks and benefits of the production season. The CSA model includes the grower estimating production costs and how many members can be supported from the production; from that information, the price of each membership or share is calculated. The produce harvested over the season is divided into each share's amount and distributed among all of the shareholder members. Each CSA is designed to meet the needs of the grower and community, and some have more than one producer and often include a variety of produce or products. CSAs can be broad or specific in the products that are available to members. Some CSAs focus on fruit crops and include raspberries; some have a CSA model that is comprised of multiple growers, with a grower providing raspberries to shareholders while other growers provide different produce or products to members. Spreading out fresh raspberry production as long as possible, such as using both primocane- and floricane-fruiting cultivars, field and high tunnel production and processed products, creates a steady supply of raspberries throughout the growing season for increased value to CSA members.

PYO marketing of raspberries includes customers traveling to the grower's farm and harvesting fruit themselves. Growers save in harvest labor, but staff members are still needed to manage visitors to the farm and facilitate their purchasing of the raspberries. PYO market success can be highly dependent on the farm's location, with the best location within 20 miles of a densely populated area (Courter, 1979, 1982; Bushway *et al.*, 2008). Any competition of PYO raspberries within the market area should be considered. Because customers harvest fewer raspberries than other PYO crops, such as strawberries, 350 PYO customers are needed to harvest 1 acre of raspberries (Bushway *et al.*, 2008). The logistics of operating a PYO enterprise for selling raspberries include determining the best parking arrangements, check-in details, transportation of customers to the raspberry fields, field supervision, containers to pick into, and promotion and communication strategies to inform customers and promote the raspberries. An attractive farm offering the necessary facilities, including parking, transportation, and restrooms, and with welcoming and competent staff is important in PYO enterprises so that the customers have a pleasant and satisfying experience and become repeat customers (Courter, 1979).

Direct-to-consumer markets bring the general public to a grower's farm to purchase raspberries or harvest and purchase PYO raspberries. It is essential for a grower to consider all risks and liability associated with the possibility of a customer being injured on their property. Even when the best possible care and precautions are taken, it is important to have adequate insurance (liability) protection (Uchtmann, 1979).

WHOLESALE MARKETS FOR RASPBERRIES

Wholesale markets include selling raspberries in fresh and/or processed form, and typically include intermediary persons involved in the sales. Examples of wholesale markets include processors, supermarket chains, restaurant chains, terminal markets, auctions and institutions such as schools or hospitals. A major wholesale market includes processors of raspberries because the perishable nature of raspberries necessitates processing to get the product to the consumer throughout the year. The majority of raspberries grown in the USA are sold fresh.

Wholesale market sales of fresh fruit to supermarkets, restaurants, terminal markets, auctions and institutions of fresh fruit include postharvest and packaging requirements to keep the shelf-life of raspberries as long as possible (see Chapter 13 for information on postharvest handling). In the short run, price change for hand-picked berries for supermarkets is due to a change in supply from the local area. Roadside market price fluctuates with supermarket prices. With raspberries in the USA, seasonal prices are due to import of raspberries. In 2011, Mexico was the largest source of US imports by a large margin. Mexico exported 16,010 tonnes to the USA (62% of US imports) (Geisler, 2012). New late-ripening autumn primocane cultivars offer raspberries out of the normal summer/early autumn season (refer to Plate 5.1); technology, such as high tunnels, can also extend the normal season into late autumn. Fresh market raspberries, sold to wholesale markets, are generally sold in single layer trays, with vented, hinged, plastic containers (clamshells) that hold ½ pint (about 125 g). The tray holds 12 one-half or 1 pint containers (USDA Economic Research Service, 1992). In New Zealand and elsewhere in the world, raspberries are sold in punnets (125 g), where two punnets equal the weight of 1 pint. Raspberries to be processed are taken to the processor in shallow containers, called lugs, which are used in the field during harvest, either by hand or mechanical means (Plate 14.7). Lugs are food-grade polyethylene, reusable and stackable.

Cook (2011) summarized the importance of shippers in the market chain for fresh produce that includes raspberries:

> 'Streamlining' the supply chain of wholesale fresh fruit has forced non-value adding costs to be removed and increased the importance of shippers in key production regions. Many shippers now act as the marketing agents for growers and are the first handler in the market chain. Larger-scale shippers also may source raspberries from many production regions to offer fresh fruit throughout the year, including shipping from several states and also other countries. Many of the shippers are specialized grower-shippers, who control a sizeable portion of fresh sales at the first-handler level.

The USA consumed approximately 122 g (0.27 lb) per capita of fresh raspberries in 2008. In that same year, there were 163 g (0.36 lb) of frozen raspberries consumed. The USA is both an importer and an exporter of raspberries. The majority of fresh raspberries consumed are imported from Mexico (Geisler, 2012).

The Washington Red Raspberry Commission (2008) in the USA estimates that 30% of processed raspberry production is individually quick-frozen, 20% is block-frozen, 30% is juiced and 20% is pureed. Given that 50% of the processed raspberries are frozen and that frozen are 36% of total US per capita consumption, it is easy to see why wholesale raspberries are an important part of raspberry production. Without wholesale, over half the market for raspberries would be lost in the USA.

Unfortunately, the costs of processing facilities are not available in the public domain. These costs will vary by individual location. The best way to determine costs for a facility would be to use an economic engineering procedure. With this procedure, the individual plant constructed first on paper and the costs at each step are determined through interviews with existing plant operators and manufacturers.

The Washington Red Raspberry Commission provides a very detailed description of the processing systems that their members follow. The following descriptions come directly from their website (http://www.red-raspberry.org/product.asp).

The Washington Red Raspberry Commission (2008) describes the current processed red raspberry products, available on the world market, as primarily red-colored raspberry fruit. Block-frozen raspberries are frozen in their own juices. Cleaned, sorted and graded berries are packed in plastic or metal pails. Standard block-frozen container weights are 2.9 kg (6.5 lb) and 12.6/13.5 kg (28/30 lb). Institutional packs are also available in 180 kg (400 lb) drums lined with plastic bagging that is 2–4 mm thick. Block-frozen raspberries may be purchased with or without sugar sweetening or syrup. Raspberries that are block frozen without sugar are referred to as 'straight pack'. Block-frozen raspberries with sugar or syrup sweetening combinations are available in ratios of 4 + 1 (4 parts red raspberries to 1 part sugar) or greater sweetening.

Individually quick-frozen (IQF) raspberries are the choicest whole frozen raspberries. 'Individual' raspberries are frozen in a quick freeze tunnel or on trays at temperatures of −20°C to −23°C (−4°F to −10°F). This 'quick freezing' seals in juices and maintains the original shape of each berry. IQF red raspberries are packed in polyethylene bags and sealed in corrugated fiber cartons. This ensures that each IQF raspberry is 'fresh frozen' and protected from damage or shipping shock. IQF raspberry packs range from 336 g (12 oz), 448 g (16 oz) and 869 g (32 oz) polyethylene bags for retail sale to 13.5 kg (30 lb) cartons. Restaurant, institutional and foodservice packs of six 2.25 kg (5 lb) IQF packages to a case are also offered.

Passing cleaned and sorted berries through a sieve to achieve a consistent particle size produces frozen raspberry puree. Screen meshes from 0.75 mm to 3.1 mm (0.03–0.125 in) determine the fineness of puree and the amount of seed removed. Raspberry puree may be processed at ambient temperatures or heated for pasteurization. Puree of red raspberries is frozen at −20°C to −23°C (−4°F to −10°F) for storage and marketing. The most common puree packs are 2.9 kg (6.5 lb) and 12.6 kg (28 lb) containers and 180 kg (400 lb) drums. Puree may be custom packed in quart and gallon equivalents. Puree is also available in concentrated form.

Raspberry concentrate is an intense capture of both raspberry essence and form. Raspberry juice is first extracted from the fruit. This juice is filtered and heated. High temperature allows the flavor and aroma (essence) to be distilled from the concentrate. The essence is captured in liquid form and may be packaged separately or mixed back into the concentrate (recapture).

Raspberry concentrate is specified in 'Brix'. Degrees brix is the approximate percent of sugar or soluble solids. Examples of red raspberry processed products and their sugar levels are presented in Table 14.1. Juice may be concentrated to a maximum level of 60–65° Brix. Raspberry concentrates are packed in 190 l (50 gal) drums and 19 l (5 gal) enamel-lined cans or 1.8, 2.25 and 2.7 kg (4, 5 and 6 lb) polyethylene-lined pails with essence packed separately or recaptured.

Traceability or the trace-back and trace-forward process is an important component of good agricultural practices (GAP) that is intended to reduce liability and prevent the occurrence of food safety problems in the marketplace (United States Food and Drug Administration, 1998). By having

Table 14.1. Examples of red raspberry processed products and their associated sugar level (degrees Brix). From: Washington Red Raspberry Commission (2008).

Processed raspberry product	° Brix
Single strength puree	8–15
Puree concentrate	20–28
Single strength juice	9.2
Juice concentrate	45, 65 and 68

a traceability system in place, raspberry growers can market fruit knowing the exact planting/field of the fruit's origin, dates of harvesting and shipping, the persons involved in harvesting and handling of the fruit, and the intended market to which the raspberries were distributed. Additional aspects may include the date and location of postharvest storage. Recently, the purpose of traceability systems has come from food safety concerns in the produce and food industries (Golan *et al.*, 2004) and some markets, such as wholesale produce, may require traceability. If contaminated raspberries occur, the contaminated product can be identified, allowing non-contaminated raspberries to be marketed (Golan *et al.*, 2004). Information that typically is required for wholesale markets includes harvest date, packing date, harvesting and handling personnel, including the staff member completing packing and quality control, and even the picker, shipping date, field identification and customer records. While traceability systems are used in wholesale fresh and processed fruit marketing, their use is also relevant and may be used effectively in direct-to-consumer sales.

Traceability systems are often linked to computer software that is designed to trace the fresh or processed fruit back to the raspberry grower or forward to the customer. Traceability codes are computerized, and there is a goal in the produce industry to adopt electronic traceability (The Produce Traceability Initiative, 2012). Barcodes are often printed on labels that are placed on the flat or container of raspberries to be marketed. Information contained in the barcode includes any aspect about the raspberries that the grower wishes to document and retain. Most traceability systems use barcodes that are read by electronic scanning machines. The electronic barcode system can be set up to print automatically, with all information stored in a computer.

Successful marketing of raspberry fruits requires intentional activities that contribute to the profitability and sustainability of a raspberry farm. Sales can include fresh or processed forms of raspberries through direct-to-consumer or wholesale markets. The market channel that will be used for quality raspberries determines the final product and its packaging, price, place of sale and promotion strategies.

REFERENCES

Bushway, L., Pritts, M. and Handley, D. (eds) (2008) *Raspberry and Blackberry Production Guide* (NRAES-35). Natural Resource, Agriculture, and Engineering Service (NRAES), Ithaca, New York.

Cook, R.L. (2011) Fundamental forces affecting U.S. fresh produce growers and marketers. *Choices* 26(4). Available at: http://www.choicesmagazine.org/choices-magazine/submitted-articles/fundamental-forces-affecting-us-fresh-produce-growers-and-marketers (accessed 13 December 2012).

Courter, J.W. (1979) Pick- your-own marketing of fruits and vegetables. *Horticulture Facts*, HM-1-79. University of Illinois Cooperative Extension Service, Urbana-Champaign, Illinois.

Courter, J.W. (1982) Establishing the trade area and potential sales for a pick-your-own strawberry farm. *Horticulture Facts*, HM-6-82. University of Illinois Cooperative Extension Service, Urbana-Champaign, Illinois.

Courter, J.W. and How, R.B. (1990) Marketing small fruits. In: Galletta, G.L. and Himelrick, D.G. (eds) *Small Fruit Crop Management*. Prentice Hall, Upper Saddle River, New Jersey, pp. 532–556.

Funt, R.C. (1990) Economics of small fruit production. In: Galletta, G.L. and Himelrick, D.G. (eds) *Small Fruit Crop Management*. Prentice Hall, Upper Saddle River, New Jersey, pp. 557–576.

Geisler, M. (2012) Raspberries. Agricultural Marketing Resource Center. Available at: http://www.agmrc.org/commodities__products/fruits/raspberries/ (accessed 23 January 2013).

Golan, E.H., Krissoff, B., Kuchler, F., Calvin, L., Nelson, K.E. and Price, G.K. (2004) Traceability in the U.S. food supply: Economic theory and industry studies. *Agricultural Economic Report*, Number 830. United States Department of Agriculture, Economic Research Service, Washington, DC.

Gradwell, S. *et al.* (1999) Community supported agriculture. Local food systems for Iowa. Iowa State University Cooperative Extension Service Bull. 1692. Iowa State University, Ames, Iowa.

Perkins-Veazie, P. (n.d.) Raspberry. Available at: http://www.ba.ars.usda.gov/hb66/121raspberry.pdf (accessed 22 February 2012).

The Produce Traceability Initiative (2012) PTI Vision. Available at: http://www.producetraceability.org (accessed 31 March 2012).

Sabota, C.M., Courter, J.W. and Archer, R. (1980) Establishing a community farmers' market. *Horticulture Facts*. HM-4-80. University of Illinois Cooperative Extension Service, Urbana-Champaign, Illinois.

Uchtmann, D. (1979) Liability and insurance for u-pick operations. *Horticulture Facts*. HM-2-79. University of Illinois Cooperative Extension Service, Urbana-Champaign, Illinois.

United States Department of Agriculture, Economic Research Service (1992) Weights, measures and conversion factors for agricultural commodities and their products. *Agricultural Handbook* 697. United State Department of Agriculture, Economic Research Service, Washington, DC.

United States Food and Drug Administration (FDA) (1998) Traceback. Section 9 in: *Guidance for Industry: Guide to Minimize Microbial Food Safety Hazards for Fresh Fruits and Vegetables.* Available at: http://www.fda.gov/Food/Guidance ComplianceRegulatoryInformation/GuidanceDocuments/ProduceandPlan Products/ucm064574.htm (accessed 26 March 2012).

Washington Red Raspberry Commission (2008) Red Raspberry Product Forms. Available at: http://www.red-raspberry.org/product.asp (accessed 22 January 2012).

15

RASPBERRY FARM MANAGEMENT AND ECONOMICS

RICHARD C. FUNT*

*Department of Horticulture and Crop Science,
The Ohio State University, USA*

INTRODUCTION

Raspberry farm management refers to the decisions that are made by the farm owner (grower) or farm manager. These decisions will be made with regards to the allocation of resources available to the grower, such as land, labor, water and capital. In many of the previous chapters, biological and technical information have been given with regards to making wise decisions about soil and site selection, water sources for irrigation, and infrastructure for the farm. For example, if a person wishes to purchase or lease (rent) land for irrigated commercial raspberry production, testing the water for high levels of iron and the soil for a high level of heavy metals would be wise. If these are high, the land should not be purchased and/or the water should not be used. The ultimate factors for a successful berry farm business are whether there will be sufficient capital to establish a crop and receipt of sufficient revenue to cover the establishment costs several years after planting, and in the final analysis, receiving a positive return on the investment over the life of the planting.

Commercial raspberry production is a business. Therefore, raspberries are considered high risk and a highly valued crop due to their high initial investment and potential for high returns, respectively, over the life of the crop. The current or potential raspberry grower/manager needs to have a large amount of information to be successful in the business of raspberry production.

Raspberries are well suited to small-scale agriculture, particularly where sufficient and efficient local labor are available. In Europe, the UK, Russia, Turkey and southern Caucasus region, raspberries have been grown for

*richardfunt@sbcglobal.net

centuries in small gardens for fresh or processed consumption (juice or beverage) and consumed at home or in nearby villages. Different types of raspberries can be harvested from spring to autumn in the field and into winter under protective cover, as hoop houses and greenhouses. As a competitive business, it is imperative to minimize costs and maximize returns. However, raspberry growers incur high fixed costs compared to other crops. The production of raspberries is labor intensive.

This chapter is written to emphasize decisions that are necessary for a successful farm business. Generally, growers are less advanced in handling business matters than with the growing of fruit (Nicholson, 1984). Therefore, an understanding of berry farm management from a business plan to a positive rate of return over the life of the planting is necessary for a satisfying experience.

While the following information may be of value to commercial growers, it may be wise for growers to organize into grower business groups and hire an agricultural economist as a consultant to advise and counsel while they learn from practical experience and plan changes in their production, management or marketing systems. A full understanding of all the input variables and their costs may not be acquired early in the enterprise. An investment in gaining information and knowledge may be the best investment a grower can make.

FARM MANAGEMENT

Raspberry farm managers are faced with a complex, multifaceted and dynamic industry. New raspberry cultivars and production systems, insect and disease management strategies, and climate change will need to be addressed. In the 21st century raspberry growers will need reliable information from many sources in order to create a business plan that leads to success. Overall, the 21st century grower lives in a global environment, global market and global economy (see Glossary 2).

Many raspberry growers have plantings of raspberries plus other crops such as vegetables and tree fruits. Biologically, the land must be well drained (internally) and fertile. The land should be located so that it can be irrigated to produce optimal yields consistently from year to year. Economically, certain high-quality soils have a greater value and provide greater profit margins than sites or soils of lower quality. In many cases, farms that sell locally grown fruits and vegetables are near metropolitan areas. However, for tax purposes, the land is zoned for agricultural production. Thus, economically, these high-value crops are considered as being produced on medium priced land, and therefore, have less risk than those on high priced land (land that could be sold for houses or industrial buildings).

When the land is first purchased, is it wise to consider cash flow in the choice of crops. Vegetable crops, such as sweet corn or pumpkins, and other crops such as wheat or rye are recommended biologically to prepare the land before planting raspberries in the eastern USA. A whole farm plan should also consider annual crops for cash flow because raspberries will take several years to produce an optimal yield.

WHOLE FARM MANAGEMENT

The whole farm budget must allocate resources that account for different enterprises and for all expenses and receipts from the different enterprises. Whole farm budgets are useful in determining the appropriate mix of enterprises to maximize returns (Castaldi and Lord, 1989). A budget leads to a plan that will utilize general equipment, refrigeration and labor effectively and efficiently. In the short term, some of the enterprises will be replaced by another enterprise and/or replanted. Thus, at the completion of a budget, a plan (or planning horizon) in the short and long term is necessary. This plan includes the hours of labor needed and whether there are areas or months where several crops have more labor required than is available.

Planting schemes need to be considered in the long term. Success of mixes of enterprises which maintain a high level of returns is brought forth early in the planning process. For example, some crops should not be replanted or rotated to the same crop land. Monoculture is neither a good soil practice nor potentially high yielding. An alternative is to use fumigation if the grower wishes to replant to the same crop. However, this practice may not be allowed due to environmental laws. Utilizing the biological and technological aspects with the economic potential is necessary under these conditions. Further, the selection of crops always requires an estimate of labor required by workers and managers. A budget and plan per enterprise is necessary for expansion or contraction of production or marketing.

ENTERPRISE BUDGET

An enterprise budget accounts for one specific enterprise, such as pick-your-own (PYO) fall-bearing red raspberries. Enterprise budgets provide the initial process (blueprint) for farm planning and analysis of the enterprise. This is particularly important when adding a new raspberry operation to the existing farm site. An example would be red raspberries that are mechanically harvested for the processing market as a new enterprise being planned in conjunction with an existing fresh PYO red raspberry enterprise (Plate 15.1).

Enterprise budgets have fixed and variable costs. Fixed costs are those costs of ownership, such as machinery, capital (principal on a loan) and interest expenses (Pritts, 2008). These costs have to be paid whether or not a raspberry crop is produced. Fixed costs vary from farm to farm as a result of farm size and choice of equipment and marketing, irrigation system, taxes and computers for record keeping. Variable costs vary directly with production and machinery usage. Variable costs include harvest supplies (containers) and activities, picking labor, mechanical harvester, transportation and field supervision of labor. It is useful to put all variable costs on a cost per hour basis. Knowing the fixed and variable costs in a raspberry enterprise can be useful in deciding whether to buy, lease or rent equipment during the non-productive years (Castaldi and Lord, 1989).

A partial budget determines the effect on net income of minor changes to an existing farm business. If a grower changed raspberry production from PYO to hand harvest for wholesale, a partial budget would be beneficial.

In 1930 in Washington County, Maryland, a survey of commercial black raspberry growers indicated an average acreage of large farms to be 5 ha (12.5 acres) to small farms (gardens) having 0.04 ha (0.1 acres). Farms having 0.8 ha (2 acres) or less had higher yields per acre per farm due, in general, to better soil selection and care given to the field. Black raspberries were grown for extra cash for the family because many men were employed in nearby towns, and thus, some farmers could be classified as part-time. However, some of these farms also grew vegetables for canneries, apples and peaches (Ross and Auchter, 1930). Most of these raspberries were sold in 21 kg (32 qt) crates and shipped to the wholesale market in Pittsburgh, Pennsylvania. Some roadside marketing was also done.

In the Maryland survey, several plantings, totaling 48.6 ha (120 acres) used the row system, a two-wire trellis and had 12,427 plants/ha (4971/acre), averaged a first-year cost of US$465/ha (US$186/acre). In the subsequent years, the average farm produced 1706 kg (2506 qt) or 78 crates. This yield is 0.7 kg/ha (32 qt/acre) or about 0.25 kg (0.78 lb) per plant. The average annual cost after the first year was US$131/0.4 ha (1 acre), harvest cost at US$1.00 per crate, container cost at US$0.55 per crate, grower, local shipping at US$0.10 per crate and US$0.70 per crate for rail transport and commission (handling at dock) at 10% of price or US$0.58/crate for a total cost of US$4.71 per crate. The gross income was US$5.78 per crate leaving a net return of US$1.07 per crate. This is a return to management and does not include family labor, if any.

RISK MANAGEMENT

Raspberry production, along with other crops and enterprises, is a high-risk industry. Risk arises from the lack of certainty. Sources of uncertainty (risk) are related to change, such as a change in the weather from extremely wet to dry (drought) to floods, which can cause lower yields (Schwab et al., 1989). Other risks, in this 21st century global economy, are changes in the world supply and demand for oil and energy sources, changes in the supply and demand for grain, declining land values in certain areas, high interest rates and labor costs (see Chapter 16). Berry growers also need to assess changes in political, social (local consumers of food and entertainment), economic and environmental circumstances where they operate.

Uncertainty can be a situation where a number of outcomes are possible; the greater the uncertainty, the greater the risk. Berry growers generally have crops that have an 8–12 year life, and therefore, must consider the future increase/decrease in interest rates, cost of labor, equipment and supplies. Creating budgets for the life of the planting and using a net present value (NPV) or internal rate of return (IRR) analysis (NPV and IRR are explained later in this chapter) can assist in managing long-term risk. Risk can be reduced when a system of culture allows the establishment costs to be covered early in the life of the planting as compared to a system that covers the establishment cost later in the life of the planting. Successful management depends on taking risks consistent with the goals and financial position of the business.

Risk management strategies for production and financial risk include diversification, i.e. growing different crops, such as winter wheat and berries, spatial dispersion (spreading crops over different climatic areas), enterprise selection, using different production and marketing schemes (PYO as well as hand or machine harvest), purchasing crop, fire and/or liability insurance, and having resource reserves sufficient to provide liquidity and cash flow for 12 months.

Managers need to take care of their own well-being and consider health, work stress and their own safety in the workplace. Reducing stress and monitoring fatigue can reduce accidents, injury and lost work time. If this is not considered, managers could lose time on the job.

The health and safety of non-hired (generally family) and hired workers is paramount to the business. Potential health hazards include hearing loss (from loud engine noise or other sources), skin cancer, respiratory diseases, and exposure to and inhalation of pesticides. For example, owners and managers should understand the need for protective measures, and insist that workers use protective clothing, including gloves and long-sleeved shirts for protection from the sun and pesticide applications.

Managers should take an inventory of worksite hazards that could be eliminated to reduce loss of work time from the job, particularly during stressful times such as planting, pruning or harvest. An inventory should include checking on power-take-off (PTO) shaft coverings, locking pesticide storage areas to keep children and other unauthorized persons away, minimizing slippery areas, and discarding unsafe ladders from which people could fall and become injured. Keep tools safe and sharp to reduce accidents and fatigue. Take breaks during the work day or create a shorter work day, if possible. In many cases a short work day can be as productive as a longer one. Safety is a habit!

MACHINES AND EQUIPMENT

Besides land, farm machines (equipment) are the single largest investment for berry growers, particularly in the USA. Growers substitute machines for manual labor especially where machines can be more efficient and cost less per hour than labor. In fact, one of the highest machine costs is the farm pickup truck.

Each operation has its own unique needs and aspirations. It is important to focus on equipment cost per unit of land (hectare or acre) and include the cost of the farm truck in overhead costs (cost spread over the entire business). In the 21st century, having new equipment with advanced technology is very important in maintaining a high level of productivity, fuel efficiency and effectiveness, as in pesticide delivery and coverage, and in completing the tasks on time. Growers should know their machinery costs and compare them with those of other growers with similar enterprises and/or with different types of equipment that may benefit the business. Berry growers who wish to maximize long-term returns will compare buying versus leasing, custom hire versus owning, investment in new versus used (pre-owned), or upgrading old to new technology. Any of these choices can either increase the cost per unit of land or reduce the cost of operating the equipment over the entire enterprise, such as more units of berries harvested and shipped per hour of use.

LABOR

Raspberries are labor intensive. There are peaks and troughs in labor needs; during the year the amount of labor can vary within a season and among years (a variable cost). Thus, a large amount of labor is needed during certain seasons. For example, pruning is performed in the spring, planting in the spring or autumn, and harvesting in late spring, summer or autumn. In extended season systems, depending on the cultivar or production system, harvest can be extended by using greenhouses or high tunnels. Examples of how technology has reduced the labor needed include using herbicides instead of hand weeding or cultivation, using mechanical pruners which can partially reduce the amount of hand pruning, and using mechanical harvesters as a substitution for hand harvested berries. Generally on small farms, the manager and/or members of the family can accomplish the pruning, irrigation and chemical sprays without hired labor. However, small raspberry growers may hire labor for fresh, hand harvested berries for roadside or wholesale markets, or for additional supervisory labor for PYO (on-farm sales) harvest.

Where there are other crops on the farm, the manager must increase or decrease the number of workers so as to get everything done in a particularly short period of time. For example, Mason and Cross (1993) found in a survey of Oregon raspberry and boysenberries growers that the harvest season was 31 days. The season in this area required over 17,500 workers to either hand harvest, machine harvest or accomplish a mix of hand or machine harvest operations (Plate 15.2).

Recruitment and retention of workers during the year can be influenced by the wage received and/or by the price received during harvest for fresh/frozen berries or processed berries which are generally harvested by machines. Also, recruitment and retention of workers can be influenced by the number of local workers. In a 1992 survey, 45% of the workers were local US citizens, 27% were aliens (both local and from other areas), 23% were non-US citizens and 4% were US citizens from other areas. Growers and managers who can retain workers year after year reduce the cost of recruitment and training. In the 1992 study, 85% of the workers returned year after year and so there was less turnaround in their workforce. Further, growers who use a mix of hand and machine harvest recognize that it takes fewer workers to harvest a unit of land than hand harvest alone.

In 1992, growers were able to manage the workforce more efficiently than in 1990 (Mason and Cross, 1993). In 1992, growers also increased the piece rate for fresh raspberries by hand harvest workers near the end of the season when berries were fewer and more scattered in the row. This increase in the piece rate retained workers, keeping them from leaving the raspberry crop and moving on to other crops. Workers were likely to leave at harvest when unharvestable, moldy berries and poor quality fruit had to be sorted from the quality berries. A large proportion of unharvested berries can be due to poor weather conditions, not enough workers to pick the entire planting at the proper time, and/or the lack of fungicide applications. In 1992, raspberry growers also hired workers for non-harvest work. Growers who provide non-harvest work are more likely to develop long-term relationships with workers. These workers can also be employed in one location and have their children attend the same school year after year (Mason and Cross, 1993).

PREPLANT TO MARKETING

Raspberry growers should first develop enterprise budgets for the pre-plant years (start-up costs), the establishment years and then for the years of production and marketing. One raspberry grower will have different costs from another grower, particularly the fixed costs. Failure to include start-up, labor, fixed and marketing costs will greatly overestimate potential profits. Yields will also vary from farm to farm and among different systems of growing raspberries. Systems are defined as the methods of growing raspberries, such as using a trellised, drip irrigated system designed for hand-harvested fresh market berries. This could be designated as an irrigated, supported system for hand-harvested, fresh market berries.

From an economic perspective, the preplant year is year zero, the planting year is year one and the next year is year two, etc. (Table 15.1). Some growers may prepare the site 2 years before planting; this could be called year minus one (−1). Yields also vary from farm to farm and from one system to another. Generally, a grower can expect 1.5 kg of fruit per linear meter (1 lb per linear ft) of row. For example, rows spaced 3 m (10 ft) apart could provide 4125 kg/ha (3639 lb/ac) of raspberries. However, many reports indicate that a yield of 2840 kg/ha (2500 lb/ac) is more accurate for black raspberries. The yield of autumn-bearing red raspberries in high tunnels may exceed 1.7 kg per linear meter (1lb per linear ft) of row (Plate 15.3) due to a longer fruiting season and larger-sized berries than those grown without protection.

ESTABLISHMENT AND HARVEST LABOR

Pre-plant costs in 2008 were estimated at US$1500/ha (US$600/acre) and included a herbicide and its application. Costs in the planting year include plants, installing irrigation, mulch, establishing sod middles, management of weeds (Plate 15.4) and pests, and an installed trellis for a total of about US$8000/ha (US$3200/acre) in the USA (NRAES-35).

Labor costs need to be part of the budget even if the grower or his family does most of the work. Generally, 8–10 h are used in pre-plant activities,

Table 15.1. Example of expense and return by year from pre-plant to first harvest.

Year	Expense	Return
−2	Removal of old plants and weeds, plus overhead costs	0
−1	Grow cover or green manure crop, plus overhead costs	0
0	Apply compost, fertilizer and lime, return from grain crop install irrigation (install high tunnel), plus overhead costs	0
+1	Plant raspberries; apply weed control, irrigated install trellis, plus overhead costs	0
+2	Prune, irrigate, control weeds, harvest costs plus overhead spray application equipment	Less than annual cost
+3	Pruning, irrigation and weed control costs, spray application equipment harvest costs (large crop), plus overhead costs and container costs (prorate if used multiple years)	Larger than annual cost
+4 to +12	Same as year 3	Equal to or greater than year 3

40–50 h in establishment (year 1) and 37.5–62.5h/ha (15–25 h/acre) in the second year when there is no harvest. Labor costs were US$10.80 per hour (including benefits) in 2010. Some raspberry systems require tying of canes, which would increase the amount of labor. In the year of harvest, harvest labor may average 2.7–3.6 kg (6–8 lb) of berries per hour. Harvest labor costs could be 40–50% of the total hours of labor per year. However, in PYO systems where the customer picks the berries, hand harvest labor costs are eliminated, but labor is still involved with greeting and exchanging money with the customers. Generally, one person will interact with 12–14 people per hour in a PYO business.

Multiple year budgeting includes a measure of the total investment and returns over the life of the planting. This could be from pre-plant to the removal of the planting after 10–12 years. The pre-plant cost is then expected to be recovered midway through the life of the planting.

NET PRESENT VALUE AND INTERNAL RATE OF RETURN

A US dollar of profit today is worth more than a US dollar of profit tomorrow due to inflation. This is known in economic terms as the 'time value of money'. The sum of adjusted US dollars over a period of years is called net present value (NPV). The NPV can be a useful tool for growers to be able to pay a rate of interest on the initial investment and allowance for risk. Generally, a discounting rate of 6–8% is used in a NPV analysis. A NPV analysis is used for a single, long-term enterprise budget. To compare two enterprises, long-term budgets that have different cost and return streams, an internal rate of return (IRR) analysis is appropriate. For example, when two crops are compared with a high and early return and another to a low and later return on investment, a decision on a long-term investment can be more accurate with an IRR analysis. Software packages are available for both of these analyses. In summary, long-term investments that utilize all labor, fixed and variable costs are essential in the management process to determine potential profits.

Failure to provide a return on equity capital in the long term for one crop means that the owner/manager did not receive a return on invested capital equal to that from alternative investment (8% on berries versus 10% on bonds). Thus, it would be more profitable for the owner/manager to invest funds in another investment than growing raspberries.

Profitability is total revenue minus fixed and variable costs. Because revenues may not occur until the second or third year (Table 15.1) after planting, growers may want to compare the payback period between crops. The payback period is the estimated time required to recover the initial costs (establishment costs or initial investment) of pre-plant and establishment costs. Thus, if one crop has a payback period of 5 years and another of

7 years, then the 5-year payback period is preferred (see Glossary 2 for definitions of economic terms).

In 1997, Funt conducted several analyses for raspberries in Ohio. First, he compared different raspberry systems with different farm sizes of 8.1, 16.2 and 32.4 ha (20, 40 and 80 acres). While the different farm sizes had different overhead costs and equipment costs spread over more acres, farm size did not influence the IRR as much as other variables. In the long term, the higher the labor requirement and start-up costs, the lower the rate of return. As an example, an autumn-bearing 'Heritage' red raspberry system with no trellis and mechanically pruned canes was compared with a summer-bearing hand-pruned red raspberry system. These systems were either hand harvested, PYO for fresh market or machine harvested for processing. The PYO autumn-bearing system had the highest rate of return over a 12-year period, generally due to a lower number of hours of labor. However, the autumn, hand-harvested system required US$1.00 more per 0.45 kg (1 lb) to make a rate of return equal to the autumn PYO system. The summer-bearing trellis PYO system was not considered to be profitable at 909 kg/ha (2000 lb/ac), receiving US$3.11/kg (US$1.40/lb). The production of 1364 kg/ha (3000 lb/ac) receiving US$2.44/kg (US$1.10/lb) was more desirable than 909 kg/ha (2000 lb/ac) receiving US$2.67/kg (US$1.20/lb). Generally, there was a 3% increase in the rate of return over the life of the planting for every US$0.22/kg (US$0.10/lb) increase in selling price (Funt et al., 1999a).

COST OF MACHINE HARVEST

For machine harvest of an autumn- or summer-bearing trellised system, lower yields and only a few hours of machine usage per year increased the cost per kilogram harvested. Under these conditions, both systems had a negative rate of return over 12 years. When a raspberry harvester is used more than 120 h per year and harvests 3031 kg/ha (3000 lb/ac), the machine harvest cost per unit was less than the hand harvest cost. However, all machine harvest berries are considered for the processing market, which generally receives a lower price per unit than fresh hand-picked berries (Funt et al., 1999b). Growers who have the management skills to produce high yields and can use a machine for 80 or more hours per year are likely to lower their harvest costs by using a mechanical raspberry harvester. Berry quality, the development of suitable markets for mechanically harvested fruit, and rates of return per unit of berries sold to fresh or processing markets, respectively, must be considered before growers can make informed decisions concerning the suitability of mechanical berry harvest for their own operations.

PRODUCTIVITY

Productivity is the key to profitability, whether it is expressed as kilograms per hour of labor or kilograms per hectare. Managers will be rewarded in the long term when many factors are considered. Productivity is correlated with the quantity of quality berries harvested. Berries that are moldy or soiled are not useable, and as explained earlier in this chapter, the harvest cost increases when moldy berries are among the clean, fresh, unblemished berries. Systems that allow for few moldy or soiled berries are preferred over those that have greater levels of spoilage. Systems that present berries at a comfortable reach for hand or PYO berries most likely have more harvested berries per hectare (acre).

If hand or mechanically harvested berries are measured by weight, then a large berry size (4 g/berry as compared to 2 g/berry) is an important economic factor in the decision-making process. Therefore, people who can pick a larger number of kilograms (pounds) per unit of land can make more dollars (income) per hour than those who harvest fewer kilograms per hour. Further, vigorous raspberry canes can produce a greater number of fruit per unit of land. Therefore, berry size and number of berries that are either hand harvested or mechanically harvested while maintaining firmness and quality are major economic factors leading to a higher level of productivity (higher amounts per hour per person). In the domestic and world market, productivity needs to increase as costs increase.

SUMMARY

The commercial raspberry grower in the 21st century will need a greater amount of information to make good decisions than was needed in the 20th century. The risk is much higher because the global market moves much faster and input costs will rise much faster than ever before. Information is needed to put many factors into a realistic plan and budget. Growers or groups of growers may require consultants on production techniques as well as consultants for the economic and business aspects. Clearly, a greater level of satisfaction will be achieved when the biological, technological and economic factors are considered together in one package.

REFERENCES

Castaldi, M. and Lord, W. (1989) Bramble crop budgeting. In: Pritts, M. and Handley, D. (eds) *Bramble Production Guide*-NRAES-35. Cornell University, Ithaca, New York, pp.131–152.

Funt, R.C., Ellis, M.A., Williams, R., Doohan, D., Scherrens, J.C. and Welty, C. (1999a) *Brambles-Production Management and Marketing*, Extension Bulletin 782. The Ohio State University, Columbus, Ohio.

Funt, R.C., Wall, T.E. and Scheerens, J.C. (1999b) *Yield, Berry Quality and Economics of Mechanically Harvest in Ohio. Fruit Crops: A Summary of Research 1998*, Research Circular 299. The Ohio State University Agricultural Research and Development Center, Wooster, Ohio, pp. 62–81.

Mason, R. and Cross, T. (1993) *Labor Demand, Recruitment and Worker Retention of the 1992 Caneberry Harvest Workforce*, Special Report 929. Oregon State University Agricultural Experiment Station, Corvallis, Oregon.

Nicholson, J.A.H. (1984) Management by Objective. *Acta Horticulturae* 155, 403–408.

Pritts, M.A. (2008) Budgeting. In: Bushway, L., Pritts, M. and Handley, D. (eds) *Raspberry and Blackberry Production Guide* (NRAES-35). Natural Resource, Agriculture, and Engineering Service (NRAES), Ithaca, New York, pp. 148–153.

Ross, H. and Auchter, E.C. (1930) *A Production and Economic Survey of the Black Raspberry Industry of Washington County, Maryland*, Bulletin No. 322. The University of Maryland Agricultural Experiment Station, College Park, Maryland, p. 208.

Schwab, G., Barnaby, G.A. and Black, J.R. (1989) Strategies for risk management. In: Smith, D.T. (ed.) *Yearbook of Agriculture*. United States Department of Agriculture, US Government Printing Office, Washington, DC, pp.151–155.

16

WORLD RASPBERRY PRODUCTION AND MARKETING: INDUSTRY CHANGES AND TRENDS FROM 1960 TO 2010

CHAIM KEMPLER[1]* AND HARVEY K. HALL[2]

[1]*Agriculture and Agri-Food Canada, British Columbia, Canada;* [2]*Shekinah Berries Ltd, Motueka, New Zealand*

INTRODUCTION

Europe is the home of commercial raspberry production, and as such, commercial and home garden raspberry production has been well established in the temperate countries of Europe for over 100 years. Archeological digs in the UK have shown evidence of raspberry cultivation since the time of the Romans, who brought the raspberry with them from Turkey. Turkey is known as the ancestral home of raspberries, being mentioned by Krataeus, Greek physician to the King of Pontus, Mithridates VI Eupator and author of a lost herbal. Agrimonia eupatorium in the first century BC mentions that red raspberries were found in Mount Ida of Frigia (Turkey) (Goodyer, 1655, in Gunter, 1934; Hummer, 2010). The species name for raspberry, *idaeus*, was adopted by Linnaeus to reflect its origin on Mount Ida in Turkey.

The United Nations Food and Agriculture Organization (FAO) world data on raspberry production begins with information on area and yields around the world for 1961 and covers the next 50 years, during which there were massive shifts in the industry. The 1960s marked the beginning of a period of political stability and prosperity around the world, which resulted in much greater wealth for many countries. At the individual household level, this period ushered in the arrival of modern conveniences and household appliances in most homes in many nations, especially household refrigerators and freezers. Up until that time, the primary household consumption of

*chaim.kempler@agr.gc.ca

raspberries was fresh during the production season and as processed, mainly as jams and jellies, for the remainder of the year. Production and consumption were limited to cool temperate regions, mostly in the northern hemisphere, with the exception of southern Australia (especially Tasmania), New Zealand, Chile and Argentina, and a limited amount in South Africa. International trade in raspberries prior to 1960 was very limited, although a significant amount of New Zealand production was sent to Australia and to the UK. All production around the world at that time was hand harvested, both for fresh consumption and for processing. Fruit quality was poor for fresh marketing and almost all fresh fruit sales were from the farm, either as gate sales or pick-your-own (PYO) sales.

Commercial production in Western countries by the 1970s had made the transition from small family blocks to larger farms where labor was hired and production for large-scale processing was the norm. In contrast, production in Eastern bloc, communist and third world countries was almost all on individual family blocks, and where production was for marketing and processing, it was accumulated at a company or collective freezing plant.

Since the 1960s we have seen enormous growth in the use of raspberries as a process product with dairy products, especially yogurt and ice cream, and for use in juices, both alone and blended with other juices. In addition, the extended ability to store as block frozen or individually quick frozen (IQF), free-flow fruit has encouraged the use of frozen raspberries in desserts, especially pies and pastries.

During the 1950s and 1960s, black raspberries in the USA were pressed for juice or made into commercial pie fillings and then into pies, which were sold by individual bakers. Fresh fruit and pie fillings were also sold to consumers for pies and consumed in the home. Today black raspberry juice or puree is used in ice cream and sold in supermarkets. Also, in research at The Ohio State University, black raspberries have been shown to reduce cancer of the colon; this could result in greater demand in the USA for fresh and processed black raspberries. In South Korea, the black raspberry (*Rubus occidentalis*) has also become widely grown and is mostly processed into black raspberry wine, a very popular health tonic in Korea.

Accompanying increases in prosperity, especially in the developed nations of Australia, New Zealand, Europe and North America, the cost of labor for harvesting raspberry fruit increased due to a diminished labor pool available for harvest. This has resulted in a significant drop in production for processing in Scotland and England; their industry is able to import frozen raspberries from underdeveloped countries. Removal of tariffs on importation of raspberries has helped this process and it has not been possible for processed raspberries to be grown there cheaply enough to compete with the imported product.

MACHINE HARVEST TECHNOLOGY

Growers in the Pacific Northwest of North America, especially in Canada where cheaper labor from Mexico is not available, have felt similar pressures. A solution to the high labor costs in this area was sought through improvements in technology, with the invention of machine harvesters that are able to shake off and collect the ripe fruit before it becomes over-mature. In the Pacific Northwest, machine harvesting is a viable solution to the high costs of labor as the cultivar grown ('Willamette') is easy to pick and suitable for this method of harvest. However, 'Meeker', introduced as a replacement for 'Willamette', is not as suitable for machine harvest. 'Meeker' produces a higher yield of larger, firmer, better quality berries than 'Willamette', but fruiting laterals are too long for effective harvest using the horizontal shake or slapper-type machines. Development of vertical shake machines has made it possible to harvest this cultivar, and because 'Meeker' is hard to pick at early stages of ripeness, but then becomes easy, the fruit samples are more uniform than with 'Willamette'. These two cultivars have been the mainstay of the Pacific Northwest process industry for many years. Since 2001, other machine-harvested cultivars have been planted in significant blocks for machine harvest and process production.

In the UK, machine harvest is not a viable option for replacing hand labor, because the cultivars grown there cannot be easily picked by machine. Additionally, canes of the cultivars grown in the UK are damaged by the passage of catcher plates, allowing pests and disease to attack the canes and causing significant plant deaths and loss of production. Development of new cultivars and of new technology for machine harvest have both been able to fulfill the requirements of a successful new cultivar for machine harvest process production in the UK, but these developments have been too late to allow the processing industry to survive or to be revived to previous UK levels.

FRESH MARKET CULTIVARS

Fresh market production around the world has expanded greatly since the 1970s and 1980s with the advent of great new cultivars and excellent proprietary production and marketing systems. 'Tulameen' was the first public domain cultivar to become extremely popular for fresh marketing. It has been followed by other new cultivars, both in Europe and North America. Sweetbriars did very well with their old cultivar 37x ('Sweetbriar') and other old cultivars, but since the company merged with Driscoll Strawberry Associates (DSA), their cultivars have excelled in the USA, Europe and other parts of the world. In the USA, production for fresh market surpassed production for processing around 2008; since then, fresh fruit has outsold the quantity of processed fruit each year.

The arrival of high-quality cultivars for the fresh market made it possible for supermarkets to get a reliable source of raspberries for consumer sales, but use of these cultivars alone has not fulfilled all the requirements of supermarkets and marketers. It has also been important for growers to learn how to supply these cultivars year-round. This need has driven the development of long cane production, staggered planting of primocane-fruiting cultivars, greenhouse (glasshouse) and tunnel-house production, and production in warm temperate, subtropical and higher altitude tropical conditions. The pursuit of reliable fruit quality has also driven production indoors, so that much of the production is now sourced from tunnels and greenhouses.

REGIONAL PRODUCTION

The FAO statistics for world raspberry production with data from 1961 have been used as the basis for the information in Figs 16.1 through 16.11, but this, in some cases, is incomplete. Some countries' production is totally or largely unrepresented, and for others it may be highly inaccurate, based on speculation rather than hard data. All the FAO data is based on information given at a national level to the United Nations from countries around the world, but some countries do not collect this information nor do they have an accurate means of collecting hard production data. For some countries, export data is available but information on the extent of local consumption is unavailable. Production figures should represent commercial production, for sales either to processors or for fresh consumption, but in some cases it appears that production figures also include home garden production and harvesting of wild stands, either of red raspberries (*Rubus idaeus*), or of other *idaeobatus* species.

Europe

In the 1960s, raspberry production in Western Europe (Figs 16.1 to 16.4) and the rest of the world was primarily for processing. At that time, there were six nations with significant production: Germany, the UK (Scotland and southern England), Hungary, Bulgaria, France and Norway. Most of these countries remain producers in Europe, but there have been some significant changes since the 1960s. One change is that raspberry production has moved from predominantly for processing in Western Europe to predominantly for fresh market. Process-grade fruit is now largely imported from developing countries outside of the European Community.

Several factors have led to the drop in production of fruit for processing in Europe. First, the costs of production have escalated, with the cost of hand harvest labor being the leader in cost increases. Secondly, imported process-grade fruit has become available cheaply due to the removal of protection for

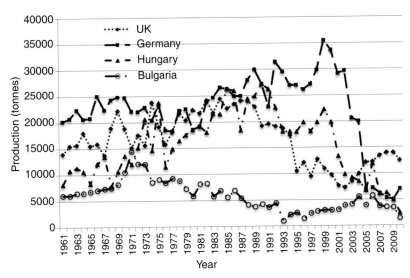

Fig. 16.1. Raspberry production in Europe (a). (Chaim Kempler, Harvey K. Hall, 2011.)

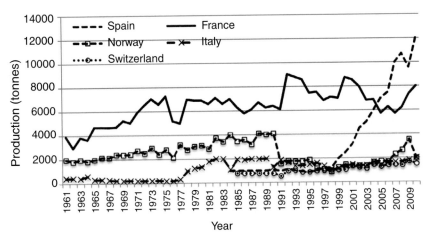

Fig. 16.2. Raspberry production in Europe (b). (Chaim Kempler, Harvey K. Hall, 2011.)

local producers and the establishment of international free trade. Thirdly, costs of production of process-grade fruit have been reduced in other countries, such as the USA and Canada, due to the use of machine harvesters. Machine harvesters have not proven viable in the UK and Europe due to problems with pests and disease associations, especially cane midge and cane blight. Damage from catching plates has been particularly bad in allowing cane midge access to attack the cane base and bring in disease. A fourth factor in the decline of raspberry production is through failure to move production and farm

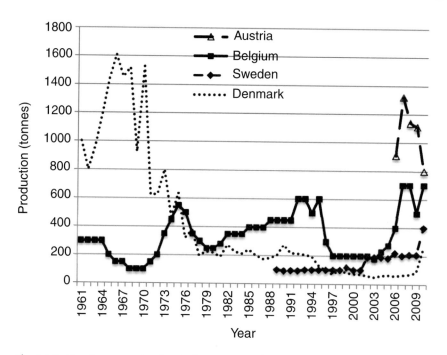

Fig. 16.3. Raspberry production in Europe (c). (Chaim Kempler, Harvey K. Hall, 2011.)

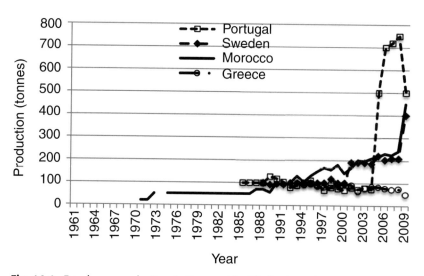

Fig. 16.4. Raspberry production in Europe (d). (Chaim Kempler, Harvey K. Hall, 2011.)

ownership to new, younger growers with the retirement and death of the post-World War II generation of farmers. Many farmers have not had family members who are keen to take on raspberry production after their retirement or death and properties producing raspberries have diminished in number.

In Europe, the decline in production was first experienced by the Netherlands and Denmark, where production of around 8000 and 1500 tons, respectively, in the 1960s, dropped to around 10% of that by the 1970s. These two countries have not recovered from this decline in production. In Belgium, production was mostly for processing in the 1960s, but by the 1970s was aimed more at the fresh market. Production in Belgium has fluctuated significantly, but in recent years has shown an increase, mainly due to tunnel production for the fresh market, driven by the excellent research carried out at Tongeren.

In the UK, a significant decline in raspberry production began in the mid- to late 1980s, resulting in the end of most of the processing production. Since 2002, production of raspberries in the UK has grown through the increase in fresh market production, mostly in tunnels or greenhouses on substrate, rather than in the soil. Two other factors also contributing to this increase in production are that most growers have joined a grower cooperative or become associated with a marketing company and superior public domain or proprietary cultivars have made marketing through supermarkets much more dependable in terms of fruit quality and promoted increased sales to consumers. Controlled marketing has enabled growers to make a reasonable profit. In the UK, raspberry research has been driven by East Malling Research (EMR) and the Scottish Crop Research Institute (SCRI) (now the James Hutton Institute), both of which have made significant investment in plant breeding and development of new cultivars. SCRI was also a center of world excellence for research in raspberries, with government-funded research into all facets of scientific investigation of raspberries, including engineering, agronomy, virology and plant pathology. Since the 1980s, many research areas have been discontinued and funding has been sought from both marketing companies and grower cooperatives, especially for the continuation of breeding programs.

In Hungary, a decline in the early 1990s was followed by a further significant drop in the first years of the new millennium, around the time Hungary joined the EU. Joining the EU put pressure on Hungarian production as the currency was exchanged for the euro. This increased the costs of production in real terms because it was no longer possible to disconnect the production costs from the market returns.

In Germany, production peaked in 1999, and by 2005 only 20% of that production remained, although consumption has grown on the basis of imports.

In 1971, Bulgarian production peaked at 14,411 tons, and thereafter declined until the late 1990s, when a modest resurgence occurred, taking production to around 5000 tons per year. France appears to have escaped the pressures of cheaper imports, with raspberry production remaining steady or gently increasing from the 1960s until the present day. This may be due to the French desire to consume locally produced fruit with superior flavor.

Production in Norway had a slow, steady increase from 1961 until 1990, when it also became subject to increased costs. By 1991, production had dropped by 60%. This has only recently seen resurgence, mostly due to the fresh market production in tunnels along the west coast fjords up to 65°N. Sweden has also been recorded as producing raspberries since the 1980s. Production has grown significantly in the 21st century, primarily tunnel-grown, fresh market fruit.

In Switzerland, production was first recorded in 1985. It has doubled since then and is all for the local, fresh market. Significant production has also been recorded in Austria since 2005, also for the fresh market.

Italian production was very low in the 1960s and 1970s, but has increased through the establishment of the grower cooperative, Sant'Orsola. This cooperative has been very successful in getting high-quality, fresh market raspberry production from many small growers and selling the fruit in Italy and all over Western Europe. Investment of EU monies has been very instrumental in getting this grower cooperative established and building the packaging, storage and shipping facilities. The only facet of this operation that could have been more effectively carried out is the distribution of grower costs, with them being shared equally overall, rather than being reduced for the large producers. Thus, several larger growers have left the cooperative and established their own marketing organizations.

Production in Spain and Portugal for export into Northern Europe during winter began in the mid-1980s and has since then grown significantly to over 10,000 tons in Spain and to over 600 tons in Portugal. This production pioneered out-of-season fruiting in tunnels, first with long canes and then with primocane-fruiting cultivars. Out-of-season production in Spain and Portugal is now a permanent part of the European market situation and supplies fruit for supermarkets throughout the period when fruit cannot be produced in the temperate Northern European winter and springtime. Production in Morocco follows production in Spain and Portugal. The supply for the European market will be sourced from North Africa for the remaining part of the year.

In December 1991, the USSR was dissolved and 15 independent states were created. In the FAO raspberry production data, this is shown clearly by the drop in production from 145,000 tons to 65,000 tons between 1989 and 1991 (Figs 16.5, 16.6 and 16.7). At the same time new production data began to be reported from Ukraine, Azerbaijan, Kyrgyzstan, Lithuania, Moldova, Latvia, Estonia and Romania. Czechoslovakia also gained its independence and separated into the Czech Republic and Slovakia. In most of the new smaller states, raspberry production has continued since the early

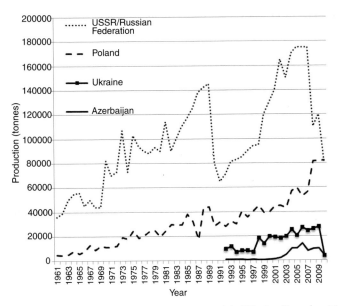

Fig. 16.5. Raspberry production in Eastern Europe (a). (Chaim Kempler, Harvey K. Hall, 2011.)

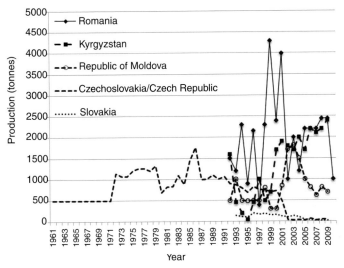

Fig. 16.6. Raspberry production in Eastern Europe (b). (Chaim Kempler, Harvey K. Hall, 2011.)

1990s, but it has frequently been unstable and many of them have suffered reduced production.

Poland also gained its independence from Communist rule in 1990, an event that showed in the production of raspberries, with a drop in production

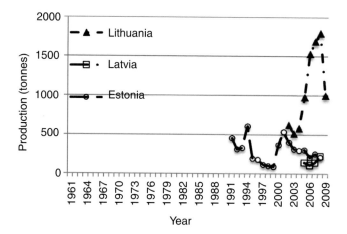

Fig. 16.7. Raspberry production in Eastern Europe (c). (Chaim Kempler, Harvey K. Hall, 2011.)

by 25% in the same year. Polish production has since advanced to over 80,000 tons and it has become an important world producer of process raspberries.

The communist government of Yugoslavia was dissolved in 1992. In the years of political instability associated with the Baltic wars of independence and the formation of new territories, much of the production has been eliminated from the new non-Serbian states, although significant production is increasing in Bosnia and Herzegovina. With the development of new cultivars in Montenegro, raspberry production is also expected to increase there. Production has varied significantly in Serbia during the years of wars, political instability and the period of sanctions. A relatively porous border has most likely given rise to increased production in neighboring countries. Production rose to a relatively stable 80,000–90,000 tons through the first decade of the new millennium (Fig. 16.8).

Raspberries are also grown in Turkey, Syria, Israel, Iraq, Iran, Afghanistan and India, but no data are available on production or cultivars grown in these countries.

Canada, USA and Mexico

Both Canada and the USA have been significant producers of raspberries since the beginning of collection of the FAO statistics in 1961 (Fig. 16.9). Black raspberry production has been confined almost entirely to the USA, both in the north-east where the fresh fruit is popular and in the north-west where production has been predominantly for processing with harvest by machine.

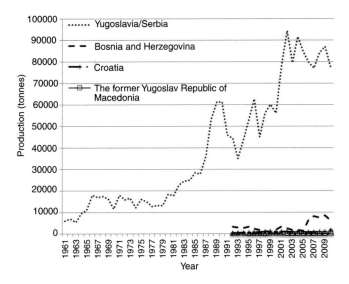

Fig. 16.8. Raspberry production in Yugoslavia and the post Yugoslav Baltic States. (Chaim Kempler, Harvey K. Hall, 2011.)

In both Canada and the USA, red raspberry production was moderate in the days prior to machine harvest. In Canada, production was between 7000 and 10,000 tons per year from 1961 to 1980, and in the USA, production has varied between 11,000 and 20,000 tons per year for this period. In the 1980s, production in Canada and the USA was nearly identical, with Canada catching up with the US production and maintaining that status for most of the decade. However, in 1989 raspberry production in the USA surged ahead and has kept increasing for the next 20 years. This increased production has been due mainly to the improved output of raspberries from California, a result of the establishment and rapid expansion of DSA (formerly Sweetbriars, Inc.) to gain up to 80% of the fresh market in the USA. Other players in the fresh market are Well Pict (on the basis of Plant Sciences cultivars), Hurst's Berry Farm and Naturipe. In the last ten years, fresh market production surpassed process production, and it continues to grow. In Canada, there has been an increase in production of fresh raspberries, particularly for sales at farm markets (C. Kempler, n.d., personal communication), and for sales in supermarkets, but production has not been sustained, with a steady decline occurring over the next 20 years.

In Mexico, raspberry production has not been a traditional activity. However, the development of raspberry production for an out-of-season supply to markets in the USA and Canada has been the driver for the development of this industry. As industry growth occurred in the USA, primarily in California, a supply was sought to be able to offer continuous,

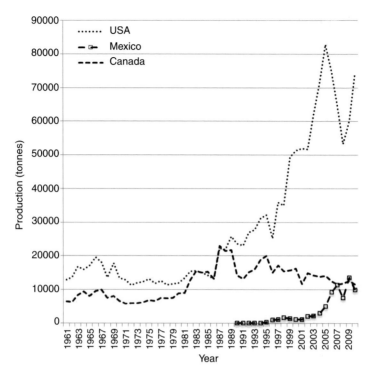

Fig. 16.9. Raspberry production in Canada, Mexico and the USA. (Chaim Kempler, Harvey K. Hall, 2011.)

year-round, fresh raspberries to supermarkets and other retailers. Therefore, the milder Mexican climate was used to achieve winter and early spring production from primocane-fruiting cultivars. In Mexico, production is achieved from cultivars such as 'Summit' and 'Joe's Baby' from DSA. The range of cultivars and the security of plantings have increased with adoption of plant protection so that proprietary cultivars can be grown in Mexico. The ways of achieving this production are different from those used for out-of-season supply in Europe, from southern Spain and Portugal to Morocco, where long cane production is used.

Fresh market quality in North America has improved dramatically since 1989 and the consumer is able to purchase fruit with a good shelf life and good eating characteristics. Slippage (losses in the market chain due to mold and other quality issues) has decreased dramatically and the profits to marketers have increased significantly, giving confidence for marketers to carry a larger stock and to devote increased shelf space to fresh raspberries.

Other Pacific Rim countries

Raspberry production in countries settled by Europeans has been practiced since the times of early settlers. In Australia and New Zealand, production for local consumption was developed, and by the turn of the 20th century, New Zealand had become the first producer of process raspberries for round-the-world international trade, with fruit being packed in barrels under sulfur. By the 1960s, this trade had diminished and much of the production in both Australia and New Zealand was consumed locally. From the beginning of the 1960s, production in Australia underwent a steady decline until the end of the 1990s when supply of fresh market raspberries became important (Fig. 16.10). Early 21st century fresh and process (machine harvest) market production increases in Australia have been primarily in Tasmania.

In the 1970s and 1980s, New Zealand production was moderate (Fig. 16.10) but it continued to supply much of the Australian demand for process fruit, which could not be provided from domestic production. During that time all process fruit sent to Australia from New Zealand was shipped block-frozen. However, this trade was not to continue as the effects of a virus disease, raspberry bushy dwarf virus (RBDV), decimated New Zealand production, which up until that time had been based on four successive RBDV-susceptible cultivars, 'Red Antwerp', 'Lloyd George', 'Marcy' and 'Skeena'. When RBDV arrived, plant health and production entered a serious decline, which resulted

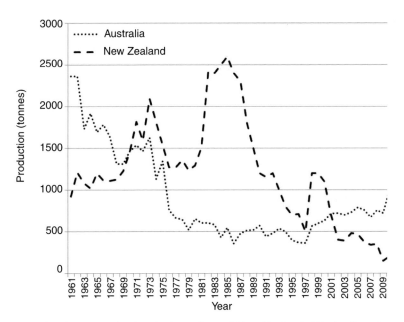

Fig. 16.10. Raspberry production in Australia and New Zealand. (Chaim Kempler, Harvey K. Hall, 2011.)

in the demise of those cultivars and the end of the export industry of New Zealand raspberries. New cultivars have been produced in New Zealand, but in 2011, none of these has been available to kick-start the processing industry. Sales of fresh market fruit from New Zealand are not permitted in Australia and local New Zealand sales are limited due to low population numbers.

In China, the Russians constructed a railway in the early 1900s, and they brought with them the tradition of growing red raspberries and the first series of cultivars, some of which are still grown in north-east China. The first main production area in China was in the Heilongjiang province, particularly around the town of Shangzhi where production for processing has been the main activity (Plate 16.1). Production for fresh markets is also growing (Plate 16.2). By 2003, 3360 mu (224 ha or 553.5 acres) were grown in China, with 3000 of these around Shangzhi. FAO data is not available for China, but production figures have been obtained for the 2006 to 2010 period (Fig. 16.11) (Hanping Dai, n.d., personal communication). In recent years, production has increased significantly and a range of new cultivars has been imported from Russia, Europe and North America. Development of new, cold hardy cultivars based on local *R. crataegifolius* germplasm which has been gathered for processing is also under way (Plate 16.3).

South Korea is the largest producer of black raspberries worldwide, but data on this production are not available from FAO statistics. Little red raspberry production is found in South Korea, but the demand for black raspberry fruit for production of black raspberry wine, popular for use as a health tonic, has increased significantly in recent years, being over 11,000 tons in 2011 (Fig. 16.11) (Lee Dong Hee, n.d., personal communication). Production is almost exclusively the North American black raspberry, *R. occidentalis*, although some consumers prefer the quality of the product produced from the local black raspberry species, *R. coreanus*. Growers are reluctant to grow the local species due to its extreme thorniness and difficulties in management.

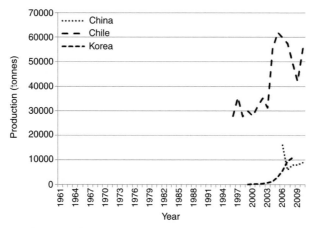

Fig. 16.11. Raspberry production in China, Korea and Chile. (Chaim Kempler, Harvey K. Hall, 2011.)

In other Asian countries, such as Japan, little raspberry production is found and climatic conditions in most countries south of China are not conducive to outdoor raspberry production.

Other southern hemisphere production

In the southern hemisphere, Chile's northern European settlers (primarily German) began raspberry production early in their South American history. This was initially for local consumption, but in the 1980s, fresh market sales to North America became a driver for increased production. When these sales started to dry up, due to cheaper production from Mexico supplying the markets, the focus on Chilean production shifted to process production, even though the cultivars grown in many cases were not suited for process use. Data is not available through the FAO statistics and information on production prior to 1996 was not obtainable (Fig. 16.11). Chile has abundant, high-quality land, plentiful water and a good climate for raspberry production, both for fresh and process markets. It is likely that Chilean production for the international processing market will grow and increase but intensive production in tunnels for fresh market (Plate 16.4) is subject to strong competition from Mexico.

In Africa, climatic conditions have not been suited to traditional cultivars of raspberries, with most climates lacking sufficient chill for raspberry production. In South Africa, small quantities of raspberries have been grown for local consumption, and only in recent years has there been an interest in growing for fresh export. Fresh raspberries have also been produced recently in the Kenyan highlands. Successful air freight to Europe and subsequent trucking to Moscow for out-of-season sales shows promise for the future of African raspberry production.

DRIVERS FOR MARKETING

Chemical components of raspberries and health properties

Raspberries contain a diverse array of flavor volatiles, nutrients, minerals and vitamins, as well as sugars and acids, which all contribute to their sweetness, aroma and health-giving properties (Table 16.1). The fruit is an excellent source of vitamins, minerals and phytochemicals that are good for human health. The composition of these chemicals can be highly variable depending on the environmental conditions, cultivar, growing location, ripeness of the fruit at harvest and storage conditions. The overall quality needs to be measured when the fruit is harvested at optimum ripeness and when it is at its highest sugar and aromatic level. The nutritional benefits are realized from the carbohydrate, vitamin, mineral, dietary fiber and polyphenolic compounds that are present.

Understanding the factors that influence consumer perception is important in the selection of new and improved varieties, in the development of new food formulations, and in the producer management practices at harvest time, postharvest handling of the fruit and treatment of the fruit after harvest.

One of the most distinguishing features of raspberries is their deep red, purple or black color, derived from anthocyanins. Differences among cultivars, ripeness and postharvest handling are some of the factors that can influence fresh or processing fruit color.

Common analyses for chemical composition of fruit, including individual sugars, total soluble solids, individual organic acids, total acidity and pH, and ascorbic acid, are indicators of fruit quality. The optimum quality characteristics as related to the appropriate harvest time to obtain maximal color and flavor are presented in Tables 16.1 and 16.2.

Year-round production

As raspberries become a fresh market product, commonly sold in supermarkets, there has been a demand for high-quality fruit to be made available year-round so that shelf space can be allocated and maintained constantly for this fruit. This has put significant pressure on the growers and marketers and has been a significant driver for modifications of growing techniques, development of specialist cultivars and movement of production into new regions. The end result has been that fresh fruit is ripening and available year-round, both in the northern and southern hemispheres.

High-quality cultivars

Modern breeding objectives include increased fruit size and yield, ease of harvest, upright growth habit, reduced spines and high yield in both primocane- and floricane-fruiting types. Successful cultivars need to be resistant or tolerant to diseases and pests, cold, heat, and drought, and adaptable to a wide range of soil types.

A successful cultivar must meet minimal criteria for many traits that are of value in the marketplace. Consumer purchase differentiation is based on potential value of the product and comparing it to similar products in the marketplace. The purchases are based on noticeable quality traits, such as flavor, color, shape, size and nutrient value of the fruit. The quality improvement in fruit quality, size and flavor with the release of the 'Tulameen' cultivar (Daubeny and Kempler, 2003) opened new markets and enhanced the demand for a fresh raspberry supply year-round.

Table 16.1. Chemical, mineral and vitamin content of red and black raspberries (data from USDA Nutrient Database).

Item	Red	Black
Water (g)	88.2	84.8
Energy (kcal)	52	64
Protein (g)	1.2	1.48
Total lipids (g)	0.65	0.80
Ash (g)	0.46	0.60
Carbohydrate (g)	11.9	14.7
Total fiber (g)	6.5	8.0
Total sugar (g)	4.42	5.44
Sucrose (g)	0.2	0.25
Glucose (g)	1.86	2.29
Fructose (g)	2.35	2.89
Calcium (mg)	25	31
Iron (mg)	0.69	0.85
Magnesium (mg)	22	27
Phosphorus (mg)	29	36
Potassium (mg)	151	186
Sodium (mg)	1	1
Zinc (mg)	0.42	0.52
Copper (mg)	0.09	0.111
Manganese (mg)	0.67	0.82
Selenium (mg)	0.2	0.2
Total ascorbic acid (mg)	26.2	32.2
Thiamin (mg)	0.03	0.04
Riboflavin (mg)	0.04	0.05
Niacin (mg)	0.6	0.74
Pantothenic acid (mg)	0.33	0.40
Vitamin B_6 (mg)	0.06	0.07
Total folate (g)	25	26
Vitamin A (IU)	33	41
Tocopherol alpha (mg)	0.87	1.07
Tocopherol beta (mg)	0.06	0.07
Tocopherol gamma (mg)	1.42	1.75
Tocopherol delta (mg)	1.04	1.28
Vitamin K (µg)	7.8	9.6

Raspberries produce new canes from the buds on the roots or from the basal buds on the older canes or the crown. On the primocane-fruiting type, flowers are initiated regardless of day length. They fruit in the current season, usually in late summer or fall. The fruiting occurs on the distal buds while the

Table 16.2. Chemical contents that affect quality characteristics of raspberries. Values are for whole fresh and processing fruit at their optimal stage of ripeness.

Item	Red (Talcott, 2007)	Black (Scheerens, J.C., 2011, personal communication)
Brix	9.26–13	7.3–11.4
Total solids (%)	14.69–17.98	18.1–18.3
Ascorbic acid (mg/100 g)	11.8–32	18.0
Citric acid (%)	1.28–1.78	0.80–0.94
Malic acid (%)	0.13–0.18	0.06–0.06
pH	2.65–3.87	3.1–4.3*
Titratable acidity (%)	1.67–2.52	0.94–1.74

*Data from Oregon Raspberry & Blackberry Commission.

proximal buds remain dormant until the following spring, potentially fruiting a second crop in the summer.

Berry crops are grown and sold via one of three marketing channels: (i) direct market through PYO or on-farm sales (grower harvested); (ii) commercial fresh market sales via local stores or shipped to more distant markets; and (iii) processed as frozen fruit, puree, dried or juice. Processed fruit may be sold directly to consumers in small retail packages, but it is often purchased by food manufacturers to make a range of products, including ice cream, yogurts, jams, jellies, juice blends, baked goods, cereals and wines.

On-farm sales and sales through farmers' markets

Standard cultivars or niche market cultivars with specialized uses or special qualities are often grown by small-scale farmers for PYO operations, for sale through on-farm shops or through farmers' markets. In the past, these growers had access to high-quality cultivars from public breeding programs, but in 2011 the best cultivars are proprietary and not available from commercial nurseries. There is a need for new, high-quality cultivars to satisfy this market and there is also an opportunity to offer the discerning consumer some cultivars with high flavor and excellent eating qualities.

Commercial fresh market

Since the 1960s, the strongest driver for changes in production and marketing has been the impetus for increased fresh raspberry sales. Older cultivars are not suitable for marketing through fruit and produce markets because the berries are soft and short-lived. Production of fruit in outdoor conditions is subject to the vagaries of weather, including rainfall, wind and intense sunshine. Development of high-quality cultivars has been the starting point of a marketing revolution during which many of the older cultivars have been shunned by marketers, retailers and consumers. Particularly in the USA, the preferred cultivars are a non-darkening, light orange-red, firm, rot resistant, long shelf-life fruit that have reduced spoilage, both in the marketplace and in the consumer's refrigerator. These cultivars are not necessarily the greatest for flavor, but the retention of the light and bright color are perceived as 'freshness' and quality, and the markets demand these fruits exclusively. As a result, sales and market share grew rapidly. Cultivars from DSA have now taken a large share of the market; other suppliers of fresh fruit have to match or exceed the quality of the DSA fruit. The DSA cultivars provide an excellent base from which to supply the market year-round, with their standard management practice of dual cropping, getting two crops a year from the same plants, cropped on the central California coast. Cropping further south in California and Mexico allows production to fill in the rest of the year.

In 2011 in the UK and Europe, the floricane-fruiting cultivar, 'Tulameen', was the cultivar of choice, as it is firm, has excellent flavor, good appearance and a longer shelf-life than the older cultivars. The light and bright color of the US varieties is not as popular there, but in recent years the DSA cultivars have also become widely grown. Other cultivars grown for the fresh market include the floricane-fruiting cultivars of 'Glen Ample' and 'Octavia'. Prior to their development, primocane-fruiting cultivars developed and available in the UK and Europe did not have the quality of 'Tulameen', 'Glen Ample' or 'Octavia'. However, newer cultivars appear to be of similar quality to both the best floricane and the DSA cultivars.

Process fruit

Process fruit is usually stored deep-frozen before being used for cooking or preparing value-added products. In the past, fruit for industrial processing was frequently stored under sulfur dioxide (SO_2), with the raspberries turning white until the SO_2 was driven off. The convenience of this process was that fruit stored in drums or barrels could be used in part, without the whole container having to be used at once. When freezing became the normal means of preserving fruit for industrial processing, the use of block-frozen fruit did not offer this utility. Unless a frozen block of fruit is cut up with a band saw, it had to be defrosted as a whole block and used at once. Another innovation for storing raspberries was the development of individually quick frozen (IQF), free-flow fruit. To originally produce IQF fruit, the fruit is frozen in single or double layers in trays and then packaged after a batch of trays has been frozen. This process was developed further to be able to produce IQF fruit by using a freezing tunnel and a liquid freezing agent, such as Freon, CO_2 or liquid nitrogen, and recently by using air coolers and 'fluid bed' technology. A modern freezing plant does not need to use a liquid coolant because a fluid bed tunnel with air blast freezers is very effective. For most commercial raspberry cultivars, it is necessary to drop the temperature of the fruit to −16°C (3°F) during the freezing process. After packaging, the temperature is further reduced, to around −20°C (−4°F). If the temperature is dropped to −20°C (−4°F) immediately, there are much greater losses through crumbling and shattering of fruit while fruit is handled over moving belts and across shakers. Before the use of freezing for preserving raspberry fruit in the home, the majority of homes used a canning or bottling process where the fruit was cooked with added sugar and then sealed in glass bottles, steel-capped sealed bottles or metal cans.

Value-added products

Raspberries are commonly used in making jams and jellies as a single product or in combination with other berries or fruits. Berries are one of the most suitable fruits for making into jams and jellies because of their quality, acidity, color, normally high pectin content, flavor and aroma. According to the US Code of Federal Regulations, the product can be made of any combination of two to five fruits, providing each of them is not less than one-fifth of the weight of the combination. Jams are a mixture of crushed (pulped) fruit or pieces of fresh fruit combined with sugar and other minor ingredients that are cooked to develop a gel involving sugars, the natural fruit pectin and the acidity from the fruit. Jellies are made from fruit juice, pectin substances, sugar and an organic acid, normally citric acid, to control acidity for obtaining the appropriate pH for gel formation. Jellies should be translucent, with very low or no pulp, forming a continuous and firm gel structure. In recent years, the production of jams and jellies has been oriented toward reducing the amount of added sugar in order to decrease their effect on the glycemic index, even developing products where the sugar has been replaced by non-glycemic sweeteners.

Dried raspberries, as fruit leather, are sold directly to consumers as a confectionary. Raspberries can also be dried to a powder and used for flavoring or added to other products for their health qualities. Freeze-dried raspberries are becoming popular as a component of 'healthy' cereal formulations or in snack bars. Raspberry juice, either alone or as a blend, is also available for healthy drinks. Other popular products made from berries include pies and other baked goods, ice cream toppings and flavorings, flavorings for yogurts, smoothies and other dairy blends and wines and liqueurs.

Machine harvest production has become the key for production of fruit for the process market in the West, with hand harvest labor costs now being too high for process production. With new cultivars, the quality of machine-harvested fruit has improved and harvesting of these for IQF production is possible, with the best of these for sales of frozen process-grade fruit. Improvements to make this possible have included easier harvest of the fruit, fruit firmness and improved quality for freezing through an IQF freezing tunnel.

REFERENCES

Daubeny, H.A. and Kempler, C. (2003) 'Tulameen' red raspberry. *Journal of the American Pomological Society* 57, 42.

Gunther, R.T. (ed.) (1934) *The Greek Herbal of Dioscorides* (1655/1933). Translated by John Goodyer. Oxford University Press, Oxford, Preface and pp. 431–432.

Hummer, K.E. (2010) *Rubus* pharmacology: antiquity to thep. HortScience 45, 1587–1591.

Talcott, S.T. (2007) Chemical components of berry fruits. In: Zhao, Y. (ed.) *Berry Fruit: Value-added Products for Health Promotion.* CRC Press, New York, pp. 51–68.

Appendix 1: Windbreaks

Michele Stanton*

The University of Cincinnati, College of Design, Art, Architecture and Planning, Cincinnati, USA

Raspberry plants are fairly sensitive to wind damage. While some air movement is beneficial to most plants, in Scotland raspberries were found to respond favorably to wind protection where average wind speed was as low as 6.1 km/h (3.8 mph) (Waister, 1970). Other studies have defined 'calm' as < 5.4 km/h (3.4 mph). If protection is desired, it is recommended that it be in place during establishment of the berry planting (Prive and Allain, 2000).

Initially, excessive wind tears, abrades or shakes unprotected plants, causing the edges of one leaf or cane to rub or strike another. This type of injury initially appears as a brown mottling (which eventually turns to purple) on upper leaf surfaces, canes and petioles (Newenhouse, 1991). Torn leaves, blossom or fruit drop, 'sandblasting' by airborne soil particles, and plant lodging may also occur.

Wind also indirectly damages plants by increasing evapotranspiration rates so that available soil moisture is rapidly depleted. Desiccation in winter can be especially problematic in areas where soil freezes. Winds may reduce the current season's crop; in strawberries, wind also reduces the next season's yields (Waister, 1973). Chronic wind exposure reduces mature plant size and fruit yield (Prive and Allain, 2000).

Acute damage can occur with infrequent high winds, such as found in passing storms. As little as 30 seconds of exposure to wind gusts of 10–20 m/s (22–44 mph) or higher are sufficient to injure plants. In either case, plants benefit from protection against wind (Hodges and Brandle, 1996).

Two main methods of protection are utilized by raspberry growers: trellises, which help stabilize plants (Vanden Heuval *et al.*, 2000) and windbreaks or shelter belts. A windbreak is a group of trees, shrubs or other plants that are installed and maintained so as to reduce and redirect wind away from a specific area. Fences composed of alternating slats or synthetic materials may also be used.

*michele.stanton@uc.edu

Windbreaks create a better environment for the planting so that plant growth and yield are higher than in unprotected areas (Plate A1.1). In some cases, windbreaks make crop production possible in areas where it is otherwise not feasible. Both air and soil temperatures in protected areas can be a few degrees warmer, especially early in the season, which may give growers a slightly earlier harvest than in adjacent fields (Brandle et al., 2004). Windbreaks can also reduce drift of chemical sprays, lessen dust and soil erosion, reduce infiltration of wind-borne inoculum, provide a more favorable habitat for pollinating and predatory insects (Vaughn and Black, 2008), and sequester more carbon in the landscape (Brandle et al., 2004). In regions where winter snowfall is a significant source of soil moisture, windbreaks help capture that moisture by decreasing drift (Brandle et al., 2004). They reduce energy needs for protected dwellings, and help create an easier environment in which to work.

Windbreaks have associated costs. In addition to the expense of installation and maintenance, the grower must consider the cost of land lost to production. The cost/benefit analysis should include land costs, crop prices, yield differential, irrigation, years until plant maturity, increased pollinator presence, decreased pesticide requirements and decreased erosion.

The height of a windbreak is its most important design factor. A windbreak protects crops on the windward side for a distance of two to five times its height, or 2H to 5H. Plants on the leeward side (away from the wind) are protected for a distance of up to 20H to 30H; protection decreases as one moves away from the windbreak, with maximum protection obtained within 10H. A single row of trees 10 m (33 ft) in height, for example, would afford protection to plants up to 200–300 m (650–980 ft) away, with the greatest degree of protection in the first 100 m (328 ft) on the leeward side. This single row should be designed without gaps, because wind can funnel through openings at greater intensity. Large berry plantings may need rows of windbreaks, which, space allowing, should be planted every 10H (Fig. A1.1).

The next design aspect is density. Medium-density windbreaks (40–60%), such as provided by a single row of native evergreen coniferous trees, are recommended for most field crops. They afford adequate protection while allowing for dissipation of turbulence. This density also gives excellent soil erosion control (Brandle and Finch, 1991). Density can be visually estimated.

The windbreak should be perpendicular to the direction of oncoming winds (Fig. A1.2). Its continuous length should be at least ten times the mature height of the tallest trees. In the case above, a row of trees 10 m (33 ft) tall would suggest that the windbreak be a minimum of 100 m (328 ft) in length. In many areas, winds may come from one direction in winter and from a different direction at another time of year. In such cases, windbreaks should be composed of multiple lines, each line at least ten times the tree height (Fig. A1.3).

Windbreaks 237

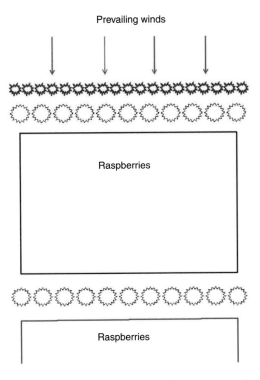

Fig. A1.1. Recurrent windbreaks for larger berry plantings. Windbreak 10 m tall repeated every 10 h (every 100 m).

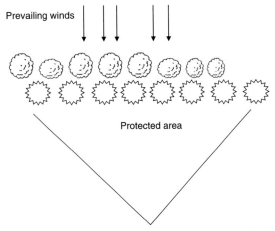

If winds come regularly from one direction, the windbreak should be perpendicular to that direction. Two rows of trees or shrubs provide better insurance against gaps.

Fig. A1.2. Single-legged windbreak with double row of trees.

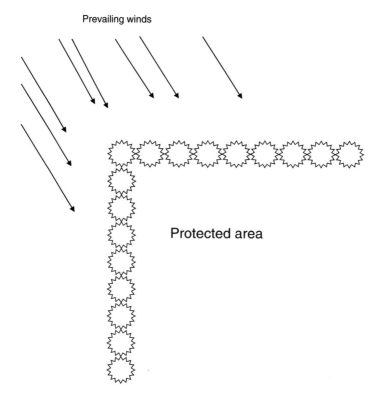

A two-legged windbreak gives better protection against wind when the planting is not at right angles to the prevailing winds, or when wind directions change with the seasons.

A single row of pine trees 2–3m (6–8ft) apart gives about 50% density.
Large deciduous species such as *Quercus* (oaks) on the same spacing may provide similar density when they are in leaf.

Fig. A1.3. Two-legged windbreak, single row of trees.

A single species of native coniferous evergreen trees is efficient; however, a mixed-species row may provide insurance against diseases or pests that might afflict a single-species planting. Species should be selected for excellent adaptability to existing climate and soils, with preference given to those that will not compete aggressively with adjacent raspberries for water and nutrients (Table A1.1).

Table A1.1. A sampling of suggested plants for windbreaks. Please keep in mind these are just a few of the large variety of plants used in windbreaks and shelter belts around the world. Trees listed here are for the tallest row of the windbreak; shrubs listed here are nitrogen fixers, and could be alternated with the trees, or used as a second row by themselves.

Botanical name	Common names	In-row spacing[a]	Mature height	Hardiness (USDA zones)	Nativity	Soil preference A, B, C, D[b]	Notes	Sources
Evergreen trees								
Casuarina cunninghamiana	River She-oak	5–8 m	20–30 m	9–10	North-eastern Australia	A, B, C, D	Takes very alkaline soils; tolerates saline soils; smog tolerant; N fixing. Can be invasive	5, 6, 14
Casuarina equisetifolia	Beach She-oak	5–8 m	14–30 m	9–10	Australia; French Polynesia, SE Asia	A, B, C, D	Takes very alkaline soils; N fixing. Can be invasive	5, 6, 14
Cupressocyparis × leylandii	Leyland Cypress	2–4 m	15 m	6–9	Wales (UK)	A	Some cultivars tolerate coastal conditions; good at odor mitigation	3, 6, 10, 11
Cupressus arizonica	Arizona Cypress	2–4 m	8 m	7–9	South-western USA (Arizona)	A, B	Takes semi-arid climates, alkaline soils	1, 2, 3, 6, 8, 14
Eucalyptus spp.	Gum, Mallee, various	5–8 m	Varies	8–10 (varies by species)	Australia	A, B	Fast growing. Seek advice for best locally adapted species. Give space; may outcompete adjacent plants for water or nutrients	3, 5, 6, 10, 14, 15
Picea abies	Norway Spruce	3–4 m	30–40 m	3–8	Cool temperate areas of Europe	A	Avoid very dry or very wet sites. Some tolerance of coastal conditions	3, 4, 6, 8, 14
Pinus eldarica	Afghan Pine	3–7 m	15 m	6–8	South central Russia	A, B	For semi-arid montane climates, alkaline soils	1, 2
Pinus nigra	Austrian Pine, European Black Pine	5–8 m	25 m	5–8	Central and southern Europe	A, B	Widely adapted; avoid in Midwestern USA. Takes acid or alkaline soils, urban conditions	1, 3, 5, 6, 8

Continued

Table A1.1. Continued

Botanical name	Common names	In-row spacing[a]	Mature height	Hardiness (USDA zones)	Nativity	Soil preference A, B, C, D[b]	Notes	Sources
Pinus ponderosa	Ponderosa Pine	5–8 m	15–30 m	5–9	Western North America; montane areas	A, B	Accepts semi-arid climates, alkaline soils. Subspecies scopulorum is very fire-resistant	1, 2, 3, 5, 6, 8, 14
Pinus radiata	Monterey Pine	5–8 m	15–30	8–10	Central western North America	A	Tolerant of trimming. Prefers coastal conditions, acid soils	5, 6, 8, 10, 14
Pinus sylvestris	Scots Pine	5–8 m	10–20 m	4–9	Northern Europe and Northern Asia	A, B	Widely adapted. Prefers acid soils. Not good in hot, humid summers	3, 4, 5, 6, 8, 14
Deciduous trees and shrubs								
Acer campestre	Hedge Maple	3–6 m	8–12 m	5–8	Europe, Western Asia	A, B	Adaptable; takes alkaline soil, some air pollution	3, 6, 14
Acer platanoides	Norway Maple	5–8 m	20 m	4–7	Eastern and central Europe; SW Asia	A, B	Widely adapted; tolerates smog. Avoid in regions where it is invasive	3, 6, 14
Acer rubrum	Red or Swamp Maple	5–8 m	25 m	3–9	Eastern North America	A, B, C, D	Takes wetter sites, smog	3, 4, 5, 6, 14
Acer saccharinum	Silver Maple	5–8 m	15–25 m	3–9	Eastern North America	A, C	Fast growing. Takes acid soils	3, 4, 5, 6, 14
Alnus rubra and other *Alnus* spp.	Red Alder, Alders	5–8 m	Varies	Varies	Species native to many areas	A, C; varies	Helps reduce pesticide drift; many are N fixing. Likes cool, humid climates	5, 6, 10, 11, 13, 14
Amorpha fruticosa (shrub)	Indigo bush, False indigo	2–4 m	2–6 m	4–9	Eastern North America	A, B	N fixing; good for erosion control	3, 13, 14
Caragana arborescens (shrub)	Siberian Peashrub	2–4 m	6 m	2–8	Manchuria, Siberia	A, B	Very adaptable; fixes N; helpful for erosion control	3, 6, 13
Ceanothus americanus	New Jersey Tea	1 m	1–2 m	4–8	Eastern North America	A, B	Nitrogen fixing	3, 14
Celtis occidentalis (many other *C.* spp. available)	Hackberry	5–8 m	7–20 m	2–9	Eastern North America	A, B, C	Widely adapted	1, 3, 5, 6, 14, 16

Crataegus monogyna	One-seed hawthorn	3–6 m	6–10 m	4–7; dislikes hot humid summers	Europe, North Africa, Western Asia	A, B	Component of traditional hedgerows in the UK. Thorny	3
Fraxinus pennsylvanica	Green Ash	5–8 m	20–25 m	Varies	Eastern North America	A	Adaptable; avoid where emerald ash borer is present. Long-lived	1, 3, 5, 6, 14
Gleditsia triacanthos inermis	Honey Locust	5–8 m	20 m	3–7	Eastern North America	A, B	Tolerant of urban conditions; does not take coastal conditions	3, 4, 5, 6, 14
Nothofagus spp.	Beech	5–8 m	18–30 m	8–10	Chile, Argentina, New Zealand	A	Takes acid soil. Likes coastal conditions	6
Populus spp.	Aspen, Cottonwood, Poplar	Varies	Varies	Varies	Species native to most temperate regions	A, C	Straight species and hybrids available. Fast-growing but shorter lived	3, 4, 5, 6, 14
Prosopis juliflora, P. spp.	Mesquite	3–6 m	8–12 m	Varies	Arid areas of North and South America	A, B	N fixer. Most like dry soil, alkaline soil. Some thorny	5, 6, 14
Quercus macrocarpa	Bur Oak	5–8 m	15–25 m	3–9	Eastern North America	A, B, C, D	Widely adapted. Some fire tolerance	1, 3, 5, 6, 14
Robinia pseudoacacia	Black Locust	3–6 m	15–20 m	4–9	Eastern North America	A, B, C, D	Widely adapted; smog tolerant; avoid in areas where now invasive	3, 4, 5, 6, 14
Salix matsudana	Corkscrew Willow	4–6 m	10 m	5–8	North-east Asia	A	Adaptable; some tolerance to drought and alkaline soils. Hybrids of this and *S. alba* can be invasive	3, 10
Salix spp.	Willow	Varies	Varies	Varies	Species native to most temperate regions	A,C	Use native species. Adapted to many conditions, including wetter sites. May have aggressive roots	3, 4, 6, 14

[a] 5–8 m is standard spacing for large trees; 3–6 m for medium trees; 2–4 m between shrubs.

[b] A, Soils that are deep to moderately deep, well drained. B, Soils that are shallow, sandy or rocky. C, Soils that are poorly drained. D, Accepts almost all soil types.

Sources: 1, Houck (n.d.); 2, Maiers and Harrington (1999); 3, Dirr (1977); 4, Ontario Stewardship (n.d.); 5, Burns and Honkala (1990); 6, Urban Forest Ecosystems Institute (2011); 7, USDA-NRCS (n.d.); 8, Earle (2011); 9, Ritchie (1988); 10, Nicholas (1988); 11, Whalley *et al.* (1985); 12, Tyndall and Colletti (2007); 13, Sun *et al.* (2008); 14, USDA (n.d.); 15, Jagger and Pender (2003); 16, Whittemore (2008).

Growers may also wish to place a second or third row of plants, often deciduous perennials or annuals, on the leeward side of the initial row of evergreen trees; these can be chosen to provide pollinator habitat, or as an additional crop for extra income. Species diversity is often helpful in providing biological control of pests or diseases (Hodges and Brandle, 2006).

REFERENCES

Brandle, J.R. and Finch, S. (1991) *How Windbreaks Work*. University of Nebraska-Lincoln Extension publication EC91-1763-B. University of Nebraska-Lincoln, Nebraska.

Brandle, J.R., Hodges, L. and Zhou, X.H. (2004) Windbreaks in North American agricultural systems. *Agroforestry Systems* 61, 65–78.

Burns, R.M. and Honkala, B.H. (1990) Silvics of North America. United States Department of Agriculture-Forest Service, Washington, DC. Database at: http://www.na.fs.fed.us/spfo/pubs/silvics_manual/table_of_contents.htm (accessed 13 December 2012).

Dirr, M.A. (1977) *Manual of Woody Landscape Plants: Their Identification, Ornamental Characteristics, Culture, Propagation and Uses*. Stipes Publishing Co., Champaign, Illinois.

Earle, C. (2011) The Gymnosperm Database. Available at: http://www.conifers.org/index.php (accessed 13 December 2012).

Hodges, L. and Brandle, J.R. (1996) Windbreaks: an important component in a plasticulture system. *HortTechnology* 6(3), 177–181.

Hodges, L. and Brandle, J. (2006) *Windbreaks for Fruit and Vegetable Crops*. University of Nebraska Extension Publication EC1779. University of Nebraska-Lincoln, Nebraska. Available at: http://elkhorn.unl.edu/epublic/live/ec1779/build/ec1779.pdf?redirected=true (accessed 13 December 2012).

Houck, M.J. (n.d.) Windbreaks Their Use. USDA-NRCS, Washington, DC. Available at: http://www.plant-materials.nrcs.usda.gov/pubs/txpmcot5584.pdf (accessed 13 December 2012).

Jagger, P. and Pender, J. (2003) The role of trees for sustainable management of less-favored lands: the case of *Eucalyptus* in Ethiopia. *Forest Policy and Economics* 5, 83–95.

Maiers, R.P. and Harrington, J.T. (1999) Windbreak tree establishment in semi-arid agricultural regions of New Mexico. *New Mexico Journal of Science* 39, 214–228.

Newenhouse, A. (1991) How to recognize wind damage to leaves of fruit crops. *HortTechnology* 1, 88–90.

Nicholas, I.D. (1988) Plantings in tropical and subtropical areas. *Agriculture, Ecosystems & Environment* 22/23, 465–482.

Ontario Stewardship, Eastern Ontario Model Forest, Ministry of Natural Resources (n.d.) *Choosing the Right Plant: A Landowner's Guide to Putting Down Roots*. Available at: http://www.ontariostewardship.org/councils/manitoulin/files/Choosing_the_Right_Tree.pdf (accessed 13 December 2012).

Prive, J.P. and Allain, N. (2000) Wind reduces growth and yield but not net leaf photosynthesis of primocane fruiting red raspberries (*Rubus idaeus* L.) in the establishment years. *Canadian Journal of Plant Science* 4, 841–847.

Ritchie, K.A. (1988) Shelterbelt plantings in semi-arid areas. *Agriculture, Ecosystems & Environment* 22/23, 425–440.

Sun, H., Tang, Y. and Xie, J. (2008) Contour hedgerow intercropping in the mountains of China: a review. *Agroforestry Systems* 73, 65–76.

Tyndall, J. and Colletti, J. (2007) Mitigating swine odor with strategically designed shelterbelt systems: a review. *Agroforestry Systems* 69, 45–65.

United States Department of Agriculture (n.d.) Fire Effects Information System Online. Available at: http://www.fs.fed.us/database/feis/ (accessed 13 December 2012).

Urban Forest Ecosystems Institute (2011) SelecTree: A Tree Selection Guide. CalPoly State University, Pomona, California. Available at: http://selectree.calpoly.edu/ (accessed 13 December 2012).

USDA-NRCS (n.d.) PLANTS Database. Available at: http://plants.usda.gov/java/ (accessed 13 December 2012).

Vanden Heuval, J.E., Sullivan, J.A. and Proctor, J.T.A. (2000) Cane stabilization improves yield of red raspberry (*Rubus idaeus* L.). *Journal of Horticultural Science* 35, 181–183.

Vaughn, M. and Black, S.H. (2008) Native pollinators: how to protect and enhance habitat for native bees. *Native Plants Journal* 9, 80–91.

Waister, P.D. (1970) Effects of shelter from wind on the growth and yield of raspberries. *Journal of Horticultural Science* 45, 435–445.

Waister, P.D. (1973) Climatic limitations on horticultural production with particular reference to Scottish conditions. *International Journal of Biometeorology* 17, 379–383.

Whalley, D.N., Jobling, J. and Marsh, P. (1985) Effects of wind and salt exposure on four hybrid cypress cultivars. *Scientia Horticulturae* 25, 93–98.

Whittemore, A.T. (2008) Exotic species of *Celtis* (*Cannabaceae*) in the flora of North America. *Journal of the Botanical Research Institute of Texas* 2, 627–632.

Appendix 2: Fertigation Systems, Calibration, Drip Irrigation and Fertigation

David S. Ross[1]* and Richard C. Funt[2]

[1]*Department of Environmental Science and Technology, University of Maryland, Maryland, USA;* [2]*Department of Horticulture and Crop Science, The Ohio State University, USA*

Fertigation is an important part of crop production and is the process of applying one or more agricultural plant nutrients through an irrigation system to the crop's root zone to meet a portion of a crop's fertilizer needs. Through this liquid feed method, nutrients may be applied as needed for maintaining good crop growth as long as irrigation water can be applied. A well-designed drip irrigation system can provide an excellent partner for utilizing fertigation with a raspberry production system in the field, a high tunnel or a greenhouse (see Fig. 9.2).

Nitrogen (N) and potassium (K)-based fertilizers are the most commonly applied nutrients by fertigation. Some formulations of phosphorus (P) and micronutrients such as zinc (Zn), copper (Cu), iron (Fe) and manganese (Mn) can also be applied if compatible with the irrigation water, particularly a pH of less than 6.5. Also, because of precipitation problems, be careful not to mix K fertilizers with calcium nitrate and Fe. For calcium nitrate and iron chelate, use stock tanks separate from those used for other fertilizer materials. Applying all P before planting, based on a soil test, is recommended (Rosen *et al.*, 2004).

Fertigation is used over the growing season during several irrigation events to apply one-half to two-thirds of the total plant needs for the season. The frequency of application is not as important as the total rate applied. Crop nutrient requirements may vary over the season and application should match those seasonal needs as a supplement to dry applied nutrients. Soil tests and leaf analyses are recommended to determine the amounts of nutrients to apply. Many local factors can affect the amount of fertilizer and water to apply.

*dsross@umd.edu

Irrigation moisture sensors should be used to determine when to apply water and for how long to maintain 50% or more of field capacity. Excessive water will leach the nutrients away from the root zone (Rosen *et al.*, 2004).

Nutrients in the irrigation pipeline can support the growth of algae in drip irrigation lines so water treatment with chlorine or other material may be needed to keep the lines clean. Fe in the water can also be a problem. Maintaining a low level of active chlorine (1 to 3 parts per million (ppm) at end of lines) with line flushing, if necessary, can reduce problems.

If the cropping is done in containers with substrates or using hydroponics, then the salts level should be monitored regularly for electrical conductivity (EC) levels so that salt levels do not climb too high. As needed, the containers should to be flushed to lower EC values.

Soil acidity and alkalinity may need to be adjusted for the crop, but fertigation is not expected to cause a problem in most cases. Iron formulation used is subject to pH levels but no toxicity of elements is to be expected (Kafkalfi and Tarchitzky, 2011).

WATER QUALITY

Water quality and fertilizer choice are the two most important considerations (Kafkalfi and Tarchitzky, 2011). Saline irrigation water is common in arid and semi-arid climatic regions and may be found in coastal areas. Sensitivity of plants to solution salinity varies between plant species and cultivars, so growers in these regions should be aware of their EC levels and plant sensitivity. Presence of nitrates (70–140 g N/m^3) and calcium (200–400 g Ca/m^3) in solution can reduce the salinity hazard to irrigated crops (Kafkalfi and Tarchitzky, 2011).

Specific ion toxicity can be a problem. Sodium (Na) in large quantities hampers root elongation and competes with Ca^{2+} on specific adsorption sites in the cell walls of the elongation zone. Chloride (Cl) is present in large quantities in saline water and can cause scorching and complete leaf death but potassium nitrate or calcium nitrate in the solution can reduce Cl uptake (Kafkalfi and Tarchitzky, 2011).

Treated wastewater (TTW) is being used in different parts of the world in increasing amounts as demand for potable water increases. TTW can add to the salt content of soils in a negative manner. However, the nutrient levels of N, P, K and micronutrients are typically higher than in fresh water. The form of the nutrient depends on any treatment process used for the TTW. The nutrient content can vary with season, source and water treatment process and the availability, amount and timing of nutrients in the TTW can vary with time. Growers who have the opportunity to use TTW should investigate the source and the treatment process, and use continuous monitoring of water quality (Kafkalfi and Tarchitzky, 2011).

For those using a micro-irrigation (drip) system, Fe, particularly ferrous iron (Fe^{2+}) with concentrations as low as 0.15–0.22 g Fe/m^3, are considered potentially hazardous for clogging drip emitters. Any water source with concentrations of more than 0.5 g Fe/m^3 should not be used for irrigation without treatment (Kafkalfi and Tarchitzky, 2011). Waters with levels of Fe above 4.0 g Fe/m^3 are considered useless for irrigation. Fe bacteria act on the Fe slime to oxidize it to form ferric iron (Fe^{3+}), which is insoluble. The Fe^{3+} surrounded by the filamentous bacterial colonies creates the sticky Fe slime gel that clogs emitters. Cl gas is a good treatment if good water and gas mixing is achieved and sand filters followed by back-up filters are used to remove precipitate (Kafkalfi and Tarchitzky, 2011).

High concentrations of Ca, magnesium (Mg) and bicarbonate (HCO_3) mean high total hardness and increase the hazard of clogging, especially when P fertilizers are introduced. Precipitation of calcium carbonate to form scale deposits is common in alkaline water that is rich in Ca and HCO_3 (Kafkalfi and Tarchitzky, 2011).

Preparing the stock solution and applying it at the correct or desired rate is important. A basic description of a dosimeter (fertigation controller) that meters a concentrated nutrient solution into irrigation water going to the crop is covered here. Maintenance of the dosimeter is critical to success, as is the regular maintenance of the equipment. Regular calibration should also be done each season to ensure proper application of fertilizer occurs (Weiler and Sailus, 1996).

DOSIMETER

The purpose of a dosimeter is to meter the concentrated fertilizer solution into the irrigation water. An injection ratio of 1:100 or 1:200 is typically used. This means that the dosimeter set at a 1:100 ratio will put 1 l (1 qt) of the concentrated stock solution into 99 liters (25 gal) of irrigation water to make 100 liters (25 gal) of dilute nutrient-rich irrigation water.

If the dilute nutrient-rich irrigation water is to contain 125 ppm nitrogen, then the stock solution (concentrate) will contain 125 ppm × 100 (proportioner ratio) = 12,500 ppm nitrogen (for a 1:100 dilution). The stock solution is made 100 times more concentrated than the final application rate so a smaller container is required and enough concentrate can be made to cover a large amount of the crop. The total amount of nutrients is dissolved into a quantity of water that matches the requirements of the dilution ratio. Care must be taken to have 100% solubility with no interaction between nutrient ingredients. The stock solution is then metered into the irrigation water at the specified ratio to make the dilute nutrient solution desired for the crop.

Different types of dosimeters are available:

1. One type is an electric motor driven pump that has a metering valve for adjusting the percentage of total flow rate that can pass through. It pumps the stock solution into the irrigation pipe after the irrigation pump.

2. Another type has an irrigation water powered pump that drives a second injection pump. This self-contained unit causes injection to occur when irrigation water flows through the pump; the injection rate is adjusted by fixed ratio settings to control the amount of concentrate added. This type of injector is available as a portable unit that can be moved around and connected into the irrigation line in the field. It is available in many small sizes for small systems and is easy to use. It takes its power from the flowing water.

3. A third type of dosimeter is a venturi device that requires a large pressure drop across it to create a suction to pull the fertilizer solution into the irrigation line. The venturi device is placed in parallel to the main irrigation line where a valve restricts the irrigation water flow to cause a pressure drop (pressure loss) to occur across the valve. The venturi device may appear to be the lowest cost system, but it may require a booster pump to create the higher pressure needed to make it work properly. The device is inexpensive but the larger pump and energy cost to operate it must be considered.

Large operators may design their irrigation systems to use a specific steady flow rate of water, say 600 l/min (lpm), for all irrigation zones. The dosimeter can be an electric motor driven piston or diaphragm pump that moves a certain lpm of stock solution, say 3.1 lpm at a 100% flow rate setting, into the pipeline. The dosimeter's injection rate can be adjusted between 1 and 100% or 0 and 3.1 gal/min (gpm) of stock solution. Thus, the dosimeter injection ratio can be varied. The actual ratio at any setting is roughly the lpm pumped by the dosimeter divided by the lpm pumped by the irrigation pump. A calibration will give the ratio.

Small operators who have irrigation zones of varying sizes may choose to use irrigation water-driven dosimeters sized for their range of water flow rates. The irrigation water flow rate through the dosimeter drives the pump injecting the fertilizer so no other power source is required. The dosimeter pump speeds up or slows down with the irrigation water flow rate and so does the rate of injection. Therefore, a constant injection ratio can be maintained over a range of water flow rates. These dosimeters are purchased in a size that matches the irrigation water flow rate, or alternatively, they can be used in a parallel circuit that carries part of the irrigation water (see Chapter 9).

CALIBRATION

Calibration is a process of checking the fertilizer injection system to see if it is functioning correctly. The goal is to verify that the desired amount of nutrients is being delivered by the irrigation system. Calibration should be done at least once per year or when any changes are made to the irrigation system or settings. A good manager will check the calibration frequently to ensure correct and economical operation. Incorrect fertigation may cause crop damage or delay in growth, which will affect crop sales.

Preparation for calibration

Several things should be checked before beginning the calibration. The equipment must be serviced and made ready in good condition. Check the suction tube strainer to see that it is clean and fluids are moving through it freely. Clean the stock tank of all settled material. Install the suction tube at least 5 cm (2 in) from the bottom of the stock tank so any solids are not pulled in. Inspect the injector and service all O-rings, if any, in the injector. Check pipe connections and any valves to see that they are properly connected and valves are open or closed as appropriate. The owner's or operator's manual for the injector should have specific instructions for maintenance and should be followed.

Calibration by volume

One method of checking the injection system is to verify that the correct amount of stock solution is being delivered into the irrigation water. This method requires a means of verifying the irrigation water flow rate. An accumulating water meter is good or for low flow rates a 20 l (5 gal) bucket may be used. Also needed is a means of measuring accurately the amount of stock solution being injected. Sizes of these containers can be varied to fit the situation.

A dosimeter ratio is determined by dividing the amount of stock solution used in liters (gallons) by the amount of irrigation water used and reduce the ratio:

$$\frac{\text{liters (gallons) stock solution}}{\text{liters (gallons) irrigation water}}$$

If the dosimeter is set for 1:100 dilution ratio, the answer should reflect that.

For example:

$$\frac{3 \text{ l stock solution}}{300 \text{ l irrigation water}}$$

The following is a simple procedure for calibrating a small dosimeter system to determine the ratio:

1. Get a measuring cup marked for 1000 ml (32 oz), a 20 l (5 gal) bucket, paper and pencil, and a calculator.
2. Measure a small quantity of stock solution into a bucket or other container from which the liquid can be removed by the dosimeter. A suggested amount would be 1000 ml (32 oz) in a measuring cup marked in ml/oz. During the test, the system will be run until the 20 l (5 gal) bucket is filled with irrigation water. If necessary, several buckets of water can be collected and dumped, just keep track of the total volume of water collected.
3. Run the injector and irrigation system for a few minutes to purge air from the injector and irrigation line.
4. Place the suction tube into the 1000 ml (32 oz) measuring cup filled to the 950 ml (32 oz) mark. Place the discharge irrigation line into a 20 l (5 gal) bucket to catch the dilute fertilizer and irrigation water solution from the injector.
5. Run the injector and irrigation system until the 20 l (5 gal) bucket is filled (or more than one bucket if necessary). Stop the injector and irrigation system at that point and remove the suction tube from the measuring cup.
6. Measure accurately the amount of stock solution remaining in the measuring cup, starting with 950 ml (32 oz) and ending with 650 ml (22 oz), for example. Subtract the remaining amount from the initial amount. In this example, 300 ml (10 oz) were injected.
7. Determine the injection ratio by comparing the amount of stock solution injected to the amount of water collected on the discharge side.

Injection ratio	= amount injected in ml (oz): amount of dilute irrigation solution discharged in ml (oz)
Injection ratio (metric units)	= 300 ml: 19,200 ml (collected from nearly full 20-l (5.3-gal) container)
	= 300 ml: 19,200 ml
	= 1:64, which is obtained by dividing both sides by 300.
Injection ratio (USA units)	= 284 g (10 oz): 19 l (5 gal) × 3.8 l (128 oz/gal)
	= 284 g (10 oz): 18.9 l (640 oz)
	= 1:64, is obtained by dividing both sides by 10.

Check the injection ratio set on the injector if it has settings. If the injector is working properly, it is set to inject at about 1:64 ratio. If the injector does not have ratio settings, then use the 1:64 ratio to create the stock solution to give you the amount of nutrient in ppm that you wish to inject.

Just to review this calculation, the stock solution was made 64 times more concentrated than the dilute solution to be applied through irrigation so the injector is mixing 1 part concentrate into each 64 parts of clear irrigation water. This results in the desired application rate of the fertilizer. In this case the injection ratio was determined by doing the calibration. Common ratios are 1:100, 1:150, 1:200, etc.

If an error occurs, check the setting of the dosimeter and perform a maintenance check on the equipment. Do the test in the same manner as normal watering so there are no differences in water flow rate or time of application. The equipment operator's manual should give instructions on making adjustments to correct the ratio.

Repeat this calibration procedure several times over the season.

8. This calibration should be repeated two or three times to verify the ratio value. Volumes of water and stock solution can be adjusted to better fit the size of equipment used and amount or flow rate of irrigation water used. The main idea is to measure both the concentrate solution and the dilute solution so the ratio can be calculated and compared to the ratio setting on the injector, if it has them. Some injectors may have a percentage total flow scale so related the ratio to the percentage setting. If the injector does not have an injection ratio, then the ratio that is determined can be used to calculate the stock solution if sufficient information is known (refer to calibration example in the next section).

Calibration by nutrient analysis

Another way to verify proper operation is to test the nutrient-enriched irrigation water delivered to the plants. In this case water samples are taken in the field and the samples are sent to a water analysis lab to test total N (nitrate, NO_3 plus ammonium, NH_4), P (ortho P and total Kjeldahl P), and K. Consider the nutrient value of the irrigation water before fertilizer is added.

For this process, knowing when the nutrients are at the sampling location is important. One way to verify this is to use an electrical conductivity (EC) meter. EC is reported in millisiemens per centimeter (mS/cm), which equals 1 mmho (millimo)/cm. **Caution:** the EC will vary for different fertilizer products, but the EC reading will show when the nutrients have reached the sampling location. This caution is for large systems where it takes several minutes for the nutrient solution to reach the far ends of a system. Sample in the middle part of the injection process to get a good sample.

The calibration procedure is as follows:

1. Start the irrigation system with clear water until the pipes are filled and the system is working.
2. Start to inject nutrients (stock solution) at the desired setting.
3. At the water sampling point in the system, monitor the EC readings every 1–2 min until the reading has peaked and three steady readings have occurred.
4. Take one or several water samples at 1-min intervals. More than one sample helps to show any variability and allows a representative average to be calculated.
5. A water sample of the clear irrigation water should be taken to establish a baseline value of the nutrients in the clear water.

6. Have a good water testing lab analyze the water samples. Test the irrigation water before fertilizer is included to have the base level.
7. Check test results against the planned values.

Parts per million

Barrett (2000) explains parts per million (ppm). Creating a stock solution means, for example, taking 'chemical A' that contains 4% active ingredient or nutrient and calculating the amount required to make 20 l at 1000 ppm concentration. Metric units are usually much easier to work with than American units and conversion to American units can be done later.

The concept of ppm can be difficult to understand, but it is a unit used to express a low concentration of a chemical in water. At a higher concentration the amount of chemical is usually expressed as a percentage.

For example 1 ppm is one red raspberry in a pile of 999,999 black raspberries (1/1,000,000). The concentration of red raspberries is 1 ppm. If there were 600 red raspberries and 999,400 black raspberries, the concentration of red raspberries would be 600 ppm. Parts per million is an expression of concentration and not an actual measure. The 600 ppm would be the same if there were 300 red raspberries and 499,700 black raspberries. One percent (1%) would be 1/100 or 10,000/1,000,000, equal to 10,000 ppm.

Stock solutions

A stock solution is a concentrated mixture of soluble dry fertilizer plus water or a purchased liquid fertilizer concentrate. A specified weight of the soluble dry fertilizer is mixed into solution in a known quantity of water to prepare a stock solution. The first thing that needs to be verified is the recipe for making the stock solution. N is a primary ingredient but two or more fertilizers may go into a recipe to supply a given mix of N, P, K and minor nutrients. Manufacturers usually supply the recipe for mixing a stock solution.

When using a dry fertilizer to make a stock solution, it is best to use a solution grade fertilizer so there are not conditioners and other materials in the dry fertilizer that might clog the injection equipment. Also, there is a limit on the amount of dry fertilizer one can dissolve in a quantity of water to make a concentrated solution. It is best to learn the solubility (g/100 ml) of the fertilizer. Read the label on the fertilizer bag to get instructions for using and mixing that fertilizer with water.

There are several ways in which errors may be introduced over time, causing the recipe to become inaccurate, so verify a recipe before use. Changes may include different suppliers, product content or bag weights. A new dosimeter may be installed with different flow rates. People may change and recipes may be understood differently. Thus, an old recipe may become a different mix by accident.

A series of mathematical calculations are required to convert grams or pounds of fertilizer into ppm of N, P or K in a stock solution. Weiler and Sailus (1996) in NRAES-56 explain how to calculate in US units. The mathematical calculations are easier in metric units. Using American units of measure one must be careful to write out all units used for conversions to be sure that the correct conversion (78 is a conversion for ounces per gallon to get ppm) is made. Units must cancel to give the final units needed. An example of calculations will be given here.

Dry (soluble) or liquid fertilizer is dissolved in water to make the stock solution. Ammonium nitrate (34-0-0) has a solubility of 18.3 g/100 ml. At 34% N, 18.3 g of the fertilizer yields 6.2 g of N. This 6.2 g of N in 100 ml is $6.2/100 \times 10,000/10,000 = 62,000/1,000,000$ or 62,000 ppm concentration. This would be a maximum concentration at 0°C (32°F). A known amount of fertilizer has been dissolved in a known amount of water, the amount of N in the solution is then calculated and then its concentration in parts per million is calculated. To apply the nutrient at some concentration, like 100 ppm or 200 ppm, the concentrate needs to be diluted by mixing (injecting) a small amount of it with a larger amount of irrigation water. This rate is typically the injection ratio of the dosimeter. In reverse, to make a stock solution to work with a fixed injection ratio, the amount of fertilizer needed to be dissolved in a known amount of water to make the concentrated ppm needs to be determined. The fertilizer bag generally gives instructions for making the stock solution (Burt et al., 1995).

The math can be tricky so the instructions on the fertilizer bag help. To apply 200 ppm N of ammonium nitrate and it is known that, for example, 64.5 mg N per 100 ml solution is to be applied, then for a dosimeter with an injection ratio of 1:100, the stock solution must be 100 times more concentrated than the solution being applied. If 64.5 mg N is the applied solution concentration, then the stock solution must be 100×64.5 mg N/100 ml or 6450 mg N/100 mg or 6.45 g N/100 ml. Note from above that this is about one-third the maximum concentration that can be made.

One can work from recommendations of kilograms (kg) per hectare (ha) and knowing the amount of water applied in irrigation in liters (l) to get a recommendation of an amount of fertilizer to apply in a volume of water. Because applications of fertilizer would be over time, the weekly amount could be used. This could be used to work out a concentration of g/l.

Making dilutions and concentrations

Barrett (2000) explains dilutions and concentrations. Fertigation involves making a concentrated stock solution or making a dilution of a concentrated fertilizer solution by using the dosimeter to mix the concentrate into the irrigation water. The equation below is handy for calculating the amount of a concentrate that is needed to make the amount of dilute solution to be applied through the irrigation system.

$$V_1 = C_2 \times V_2 / C_1$$

where:

V_1 is the volume or amount of original chemical needed
C_1 is the concentration of the original chemical
C_2 is the concentration desired for the solution being made
V_2 is the volume of the second solution being made

In using this equation, the concentrations for C_1 and C_2 must be expressed either in ppm or as a percentage. Likewise, the units for volume must be the same: liters, quarts, gallons, etc.

Note that the concentration of 'chemical A' is 4%, while the concentration of the final solution is given as 1000 ppm. Therefore, converting the 4% to ppm gives 40,000 ppm, from 1% = 10,000 ppm.

Next, the amount of 'chemical A' needed to make 20 l at a concentration of 1000 ppm is to be determined. Given that C_1 is 40,000 ppm, C_2 is 1000 ppm and V_2 is 20 l:

$$V_1 = C_2 \times V_2 / C_1$$
$$V_1 = 1000 \times 20 / 40,000$$
$$V_1 = 0.5 \, l$$

When using a dosimeter, a dilution ratio of 1000 ppm/40,000 ppm would be used. This would be 1/40 setting. One unit of concentrate (at 40,000 ppm) would be applied in 40 units of irrigation water to deliver 1000 ppm.

Calculations *by weight* can also be done using the equation where weight is exchanged for volume:

$$W_1 = C_2 \times W_2 / C_1$$

Taking a dry 'chemical B' that contains 20% of a particular active ingredient or element to make 50 kg with a concentration of 0.5% (C_1 is 20% (200,000 ppm), C_2 is 0.5% (5000 ppm) and W_2 is 50 kg):

$$W_1 = C_2 \times W_2 / C_1$$
$$W_1 = 0.5 \times 50 / 20$$
$$W_1 = 1.25 \, kg$$

In other words, to make the lower concentration, use 1.25 kg of the active 20% material in 50 kg of final mix for a 0.5% (5000 ppm) concentration.

Manufacturers may have charts or tables for their product to help make the stock solutions and for calculating dilutions.

Calculation for making a stock solution

Boyle (2003) explains the calculation of a stock solution. A stock solution is a concentrated solution of fertilizer made by dissolving fertilizer in water that is intended to be injected into the irrigation water by a dosimeter. By mixing with the irrigation water, the concentrated solution is diluted to the rate at which the nutrient is to be applied. Thus, knowledge of the dilution ratio (injector ratio) of the dosimeter must be known. Also, one must know the application rate of the fertilizer in terms of ppm of N, or P or K. The rate might be 100 ppm N or 200 ppm N, for example. An injection ratio of 1:200 means that the stock solution will be made 200 times more concentrated than the rate at which the nutrient is applied. A stock solution for application of 150 ppm N by an injector with a dilution rate of 1:200 would mean the stock solution would be 200 × 150 = 30,000 ppm N. The goal is to determine the weight of fertilizer to dissolve in a volume of water to make the stock solution.

The basic equation for this calculation is:

$$\text{Amount of fertilizer to make 1 volume of stock solution} = \frac{\text{Desired concentration in parts per million} \times \text{Dilution factor}}{\text{\% of element in fertilizer} \times \text{Constant, C}}$$

where the dilution factor is the larger number of the fertigation controller injection ratio and the conversion constant C is determined by the units desired:

Unit	Conversion constant, C
Ounces per US gallon	75
Pounds per US gallon	1200
Grams per liter	10

Example 1. A grower has a 1:200 fertigation controller and a fertilizer with an analysis of 15-16-17 (%N-%P_2O_5-%K_2O). He wants to apply a 100 ppm solution of N at each watering. How many grams of fertilizer would he have to weigh out to make 1 l of concentrate?

List all variables:

1. Desired concentration in parts per million (ppm) at application = 100.
2. Dosimeter ratio = 1:200; dilution factor = 200.
3. Fertilizer analysis = 15-16-17 (15% N).
4. Grams of fertilizer to make 1 liter of concentrate = X (unknown); use 10 as the conversion factor C.

Set up and solve the problem using the equation:

$$X = \frac{100 \text{ ppm N} \times 200}{15\% \text{ N} (15) \times 10} = \frac{20{,}000}{150} = 133.33 \text{ g} = 0.13 \text{ kg} \quad (0.13 \text{ kg/l})$$

Add 133 g/l of water to create this stock solution. Larger quantities can be made by increasing the amount of fertilizer times the number of liters of water.

Equations can be used to make various calculations based on what one knows and what one wishes to find. Also, many fertilizer companies have charts that give the amount of fertilizer to use for each of several injection ratios.

N-P$_2$O$_5$-K$_2$O versus NPK

If the calculation is for mixing actual P or K then an adjustment is needed when using the fertilizer analysis. Government regulations in the USA require content to be listed as %N: %P$_2$O$_5$: %K$_2$O. However, plants respond to N, P, K.

$P_2O_5 = 43.7\%$ P

$K_2O = 83\%$ K

Actual P available is only 43.7% of the amount of P$_2$O$_5$ in the fertilizer and actual K is 83% of the K$_2$O in the fertilizer. An additional calculation would be required if actual P or K is to be determined.

Recommendations are often given in terms of N, but one can consider the amount of P and K being delivered by a fertilizer product.

Water pipe flow rates (velocities) (and time for chemical to reach end of the irrigation line)

Understanding water flow rates in the main line distribution system and in manifolds or submains is important when doing a calibration test, particularly when collecting samples to test the EC at the end of an irrigation line. It may take some time for water to reach a given location in the system and samples should not be drawn until the nutrient solution has reached the sample site.

The water flow rate, Q, is explained by the following relationship:

$Q = A \times V$ ($\times$ conversion units)

where Q = water flow rate, gallons per minute (gpm)
$\quad A$ = cross-sectional area of pipe, square feet
$\quad V$ = water velocity in pipe, feet per second (fps)

Basically, this relationship says that for a given flow rate, if the pipe size (A) changes, the water velocity (V) must change in the opposite direction. If the pipe gets smaller, the water velocity must get larger, in order to get the same amount of water through the pipe. Also, if the flow rate reduces because water is being discharged from the manifold into many laterals and the manifold stays the same size, then the water velocity must decrease past each lateral.

The equation above and its relationships can be read on a friction loss chart that shows the water velocity and friction loss (an energy loss in pressure) for various flow rates in different size pipes. A dealer can provide this chart for all types and sizes of pipe and explain the relationships. In designing a system, maintaining the pressure to the end of the system requires careful selection of all pipe sizes. A dealer uses the friction loss chart for this purpose and it can be useful to the grower also.

The travel time for a chemical from the injection point to the final application point can be calculated through each pipe segment using the following equation and then adding all the times for all segments. For one pipe segment:

$$\text{Time} = \frac{\text{distance}}{\text{velocity}} = \text{distance} \times \frac{\text{inside area of pipe segment}}{\text{flow rate through the segment}}$$

where: time = minutes
         distance = pipe segment length, feet
         inside area = square feet
         flow rate = gallons per minute, gpm

BACKFLOW PREVENTION TO PROTECT DRINKING WATER

When a potable (drinking) water source is used for irrigation, particularly where fertilizers or other chemicals are added to the water, it is very important to avoid a cross-connection that might allow the chemical to flow back into the potable water source. This would be a backflow of the water to contaminate the source. A backflow could happen if there is a siphoning effect due to a low pressure on the source side of the connection. A hose connected to a faucet would be a cross-connection. An air gap device that opens at low pressure can prevent that.

Irrigation systems are typically under pressure, so check that valves, which close if the water tries to flow in the reverse direction, are used to prevent backflow. There are different types of and different levels of protection provided by different backflow preventers.

At a minimum, an atmospheric vacuum breaker (AVB) or pressure vacuum breaker (PVB) is required. Both of these must be located at a certain point above the highest point in the irrigation system. Both only protect against back siphonage. The AVB opens to the atmosphere to create an air gap to break the siphonage, while the PVB opens to dump the flow of water that was under pressure. A double-check-valve assembly (DCA) protects against both back pressure and back siphonage. A DCA is allowed in some US states. The reduced pressure (RP) type backflow preventer is most expensive and protects against both back pressure and back siphonage. It will dump large quantities of water when tripped (Vinchesi, 1999). An agricultural manufacturer is developing a unit with an AVB and a check valve in a backflow preventer for low hazard situations to protect irrigation wells (non-potable) and ponds.

RASPBERRY NUTRIENT RECOMMENDATIONS

As indicated at the beginning of this appendix, drip irrigation is not just for applying water to raspberries. Using a drip irrigation system to deliver nutrients directly to the roots can be a very economical method of improving yields while reducing material and labor costs. In general, drip irrigation is used in the humid eastern USA from mid-spring through the summer months, ending in early summer for floricane-fruiting raspberries. Primocane-fruiting raspberries will flower and fruit in mid-summer to autumn and until the first frost. Applying N and P to raspberries at critical growing stages can improve primocane and floricane growth and improve the number of flower buds, firmness and size of berries.

Fertigation is a term for a process for injecting one or more nutrients through an irrigation system using a dosimeter to deliver a quantity of nutrients only to the root zone of plants. Drip irrigation experiments in New South Wales (Black, 1971) applied water to 50% of the root zone and accounted for 90% of the trees' need for water. In Maryland, blueberries with one line of drip irrigation and sawdust mulch had the same number of shoots and length of shoots as two lines of drip (Funt et al., 1979). In New York, black raspberries that were treated for three seasons with sulfate of potash at the rate of 227 g (8 oz) per plant increased yield by 20%, number of canes by 25% and had a 100% increase in pruning weight (Tomkins and Boynton, 1959). Many studies indicate that drip irrigation, especially when water is applied only to the root zone and not the row middles, can reduce the amount of water needed per hectare (acre) by 50%, that N can be reduced up to 50% when applied directly to 50% of the root zone, especially when N is fed over time so that little is leached as compared to broadcasting dry fertilizer. Typically, drip irrigation applies water to a limited surface area above or within the root zone so the amount of applied water is much less than is applied by overhead broadcast application. Therefore, there are lower costs for water, fertilizer, labor and farm equipment (tractor and fertilizer spreader) as compared to conventional systems. In Ohio, experiments with high bush blueberries on raised beds having a soil organic matter at 2.5% indicated the same leaf N, number of canes, cane height, yield, berry size and pruning weight for applications of 22 kg/ha and 44 kg/ha (20 and 40 lb/acre) of N injected as a split application as compared to a split application of 60 lb/acre of actual N spread by machine (Funt and Bierman, 2000).

Supplemental fertigation in fields

Raspberry fertilization can follow the instructions given here but nutrient testing (soil or leaf (tissue) testing) is recommended to account for different soils and plant responses. Some nutrients are incorporated into the soil before planting (refer to Chapters 8 and 12) and fertigation is used to supplement plant requirements over the growing season. In the planting year (year 1 – first leaf), transplants are set in soil which has compost or manure and nutrients such as Ca, P, K and minor elements that were applied in the pre-plant year (year 0). At planting transplants are fertilized in early spring with a transplant solution and followed by 5.5–11 kg/ha (5–10 lb/acre) of actual N, 2–4 weeks after planting. Beginning 8 weeks after planting and for the next 8 weeks, 0.5–1.0 kg/ha (1–2 lb/acre) of actual N can be injected per week for an additional 8 weeks. Take leaf samples during week three and adjust the rate of N. More N may be necessary in sandy soils and/or soils having less than 2% organic matter.

In the second and subsequent fruiting years, after plants are 5–10 cm (2–4 in) tall in the spring, begin injection for 8 weeks at the rate of 2–3 kg/ha (4–6 lb/acre) of actual N and approximately the same amount of K, if K levels are medium to low. Potassium nitrate can be a good source of N and may be supplemented with calcium nitrate in another stock solution (**caution**: do not mix in same stock solution). In regions where P is low, phosphoric materials may be injected alone at the beginning of the season and then dropped out of the stock solution during the growing season. The same amount of N should be continued for 8 more weeks for a total of 16 weeks. Do not over-water or fertilize floricane raspberry plants in late summer to allow acclimation to occur normally in cold regions. For autumn-bearing primocane production, plants should receive the higher rate of N beginning 3–4 weeks before the first harvest (10% of total crop). Further, maintain optimal soil moisture 3–4 weeks before harvest and during the first 6 weeks of harvest for optimal fruit size. Monitor soil moisture with moisture sensors.

Fertigation in high tunnels

In high tunnel production, the soil should be well prepared before planting by incorporating compost, P, K, N (just before or after planting) and minor elements in the optimal range for raspberries based on soil tests, including pH testing. Maintaining optimal soil moisture will be most critical for the entire season for root and cane development because plants will not receive water from rainfall. Therefore, the injection of nutrients may be more beneficial on a daily basis rather than a weekly basis. The primary rationale is to supply the amount of actual N over the season. Because plants will be in a warmer temperature regime, plant growth and development will be more rapid inside than outside of the high tunnel. Secondly, for autumn-fruiting primocanes that ripen in late autumn the injection of N may be continued up to and through the second harvest (20% of the entire crop). When N is supplied by injection, levels in the soil and plant should remain high for 3 weeks after the injection of N is stopped.

For organic production in high tunnels, synthetic fertilizers are not allowed. Composted chicken or dairy manure is preferred for their N levels. Grass or leaf composts have also shown some good results, with grass being high in N and leaves having a high level of Ca. These are particularly good incorporated 10–15 cm (4–6 in) into the soil in the row before planting on either flat or raised beds. For poultry or dairy composted manure, the N levels should be assessed to determine the amount of manure required to supply N equal to 132–220 kg/ha (60–100 lb/acre); these are placed on top of the soil around the plants. Composted manures are preferred because they have very low weed seed germination as compared to fresh manures.

Fertigation for greenhouse raspberries in containers

Raspberry plants (tissue culture virus indexed transplants are best) are planted into 12-l pots (3 gal), set into soilless medium (standard peat, vermiculite and sand in a 2-2-1 ratio) with calcium carbonate (lime), P and N fertilizer plus micronutrients and set outdoors after the first frost for the first growing season. Plants are irrigated daily (more often when the weather is hot) and 100 ppm of N is injected once per week. In early fall, water and fertilizer are reduced significantly, but pots should not be allowed to become excessively dry. This acclimation process of 1200 hours at −2°C to +6°C is necessary for flowering and fruiting. Plants are moved into a cooler for 8 weeks so that dormancy is completed by early winter. Another method is to buy dormant plants in early winter, pot them and set pots immediately into the greenhouse (Demchak, 2008). Studies in Canada on several cultivars of red raspberries applied 40 g (1.5 oz) of 100-day slow release fertilizer (14N-14P-14K) per 4-l (1-gal) pot with drip irrigation (water only) in May 1997 followed by the same amount in January 1998 and then 50 g 13-13-13 100-day slow release with micronutrients in December 1998 (Dale et al., 2001). In this study of different red primocane cultivars, the yield varied among cultivars but the amount of fertilizer applied provided good cane growth and demonstrated that raspberries could be grown with this fertilizer and drip irrigation in the greenhouse throughout the year and have good yields.

Summer-grown potted plants may be moved into the greenhouse and given 50 ppm N per week plus a soluble fertilizer of 5-11-26 (% of N-P-K) (separate injection tank). Once moved into the greenhouse, flowering and fruiting begins over an 8-week period. Fertigation may be regulated to lower amounts if first fruits become soft. Adding additional K (as potassium nitrate) may increase fruit color and firmness. After reaching frost-free days, the plants are then moved outdoors and the cycle is repeated. Placing moisture sensors in a few pots will be useful to maintain optimal moisture and oxygen levels in the root zone.

REFERENCES

Barrett, J. (2000) Getting comfortable with PPM, percentages and metrics. *Greenhouse Product News* (GPN), May 2000, pp. 20.

Black, J.D.F. (1971) *Daily Flow Irrigation*. Leaflet 191. Department of Agriculture, Victoria, Australia.

Boyle, T.H. (2003) *Fertilizer Calculations for Greenhouse Crops*. University of Massachusetts Extension Greenhouse Crops & Floriculture Program. Available at: http://extension.umass.edu/floriculture/fact-sheets/fertilizer-calculations-greenhouse-crops (accessed 13 December 2012).

Burt, C., O'Connor, K. and Ruehr, T. (1995) *Fertigation*. The Irrigation Training & Research Center, California Polytechnic State University, San Luis Obispo, California, pp. 91–102.

Dale, A., Gilley, A. and Kent, E.M. (2001) Performance of primocane-fruiting raspberries grown in the greenhouse. *Journal American Pomological Society* 55, 27–33.

Demchak, K. (2008) Production methods In: Bushway, L., Pitts, M. and Handley, D. (eds) *Raspberry & Blackberry Production Guide*. NRAES, Ithaca, New York, pp. 28–38.

Funt, R.C. and Bierman, P.R. (2000) Subsurface drainage, raised beds and nitrogen fertigation management for blueberries. Department of Horticulture and OSU Centers at Piketon, The Ohio State University. Unpublished report.

Funt, R.C., Ross, D.S. and Brodie, H.L. (1979) *Evaluation of Trickle Irrigation in Maryland*. Maryland Agricultural Experiment Station, Agricultural Engineering Department, College Park, Maryland. No. A2662 Scientific Article, Contribution No. 5703.

Kafkalfi, U. and Tarchitzky, J. (2011) *Fertigation – A Tool for Efficient Fertilizer and Water Management*. International Fertilizer Industry Association, Paris, pp. 63–73.

Rosen, C., Wright, J., Nennich, T. and Wildung, D. (2004) Fertility and fertigation management – high tunnel production. *Minnesota High Tunnel Production Manual for Commercial Growers*. Section 8 of M1218. University of Minnesota, Minneapolis/St Paul, Minnesota.

Tomkins, J.P. and Boynton, D. (1959) A response to potassium by the black raspberry. New York State Agricultural Experiment Station, Cornell University, Geneva, New York. Journal Paper 1149, pp. 1149.

Vinchesi, B.E. (1999) Backflow basics. *Landscape & Irrigation* February, 48, 49, 52.

Weiler, T.C. and Sailus, M. (ed.) (1996) *Water and Nutrient Management for Greenhouses*. NRAES-56. Natural Resource, Agriculture and Engineering Service, Cooperative Extension, Ithaca, New York. Available at: http: www.nraes.org.

GLOSSARY 1: BIOLOGICAL TERMS

Acclimation: the natural process of plants adapting to a climate; hardening; the phase under declining hours of sunlight and temperatures in late autumn, when shoots stop elongation and tissues acquire increased cold hardiness for winter.
Airblast sprayer: a machine using a pump, specially designed nozzles and a large fan to deliver mixtures of pesticides and/or nutrients to a plant canopy in order to reduce pest damage and to improve plant growth, fruit size and fruit quality.
Anthocyanin: water-soluble pigments that may appear red, purple, or blue according to the pH. They are flavonoids and are odorless and nearly flavorless.
Antioxidant: a substance that opposes oxidation or inhibits reactions promoted by oxygen or peroxides; raspberries may contain antioxidants, as calcium, potassium, magnesium or selenium and other compounds; may be expressed in total amount in an ORAC analysis.
Axil: the angle between a petiole and the stem to which it is attached.
Axillary buds: buds that develop in the leaf axils of a primocane, which may break bud and begin growing vegetatively if the growing apex of the plant has been removed.
***Botrytis cinerea*:** the fungus responsible for gray mold of fruit, one of the most serious diseases of raspberry throughout the world.
Brix: a measure of a fluid's total soluble solids concentration; expressed in degrees or as a percentage of weight of sugar in the solution.
Calyx: the five green petal-like projections beneath a flower; usually seen as a cap attached to the top of the fruit just beneath the stem.
Cane: a main stem of a small fruit plant; also a woody, mature shoot after leaf fall.
Cation: generally a positively charged ion.
Cation exchange capacity (CEC): the maximum quantity of total cations, of any class, that a soil is capable of holding, at a given pH value, for exchanging with the soil solution. CEC is used as a measure of fertility and nutrient retention capacity.
Certified stock of raspberries: plant material certified as free of pests and disease. It is sold as Foundation grade plants to nurseries.
Chilling requirement: number of hours at or below a certain maximum temperature that is necessary to produce internal changes in a plant that result in uniform bud break and the normal sequence of growth following winter dormancy; plant needs to be exposed to temperatures between 0 and 7°C (32–45°F) once it is dormant.

Chilling unit: period of time (usually 1 h) at or below a specific threshold temperature that has the maximum effect toward fulfilling the chilling requirement of a given plant.

Clay: smallest soil particle class size, less than 0.002 mm in diameter; so small they cannot be seen except with an electron microscope.

Climacteric fruits: fruits that have high respiration rate during fruit ripening, a process mediated by the phytohormone, ethylene, which dramatically increases up to 1000 times of the basal ethylene level as ripening occurs. Raspberries are non-climacteric fruits.

Dormancy: a plant's physiological stage of rest; usually environmentally induced as nights lengthen in fall and temperatures drop below 7°C (45°F).

Dosimeter: in agriculture, an instrument that measures fertilizer and pesticides; also referred to as a fertigation controller, injector, proportioner.

Drip irrigation: refers to the application of water to crops in small amounts, under low pressure and applied directly to the root zone under low pressure; also referred to as microirrigation or trickle irrigation.

Drupe: fruit derived entirely from an ovary, one seeded with a pericarp, fleshy mesocarp and stony endocarp.

Drupelet: any of the small individual drupes forming a fleshy aggregate fruit, such as a blackberry or raspberry.

Effective neutralizing value (ENV): a measurement of the effectiveness of the particular quality of lime for agricultural use. The ENV of a lime is the ability of a unit mass of lime to change soil pH.

ELISA (enzyme-linked immunosorbent assay): a common serological test for particular and unique antigens or antibodies that have been produced from viruses or microbes attacking a plant. Detection of a response to an antibody produced to a specific virus is a clear indication of the presence of a virus.

Evapotranspiration (ET): the amount of water lost from the soil surrounding a plant plus the water lost by transpiration from the plant itself.

Fertigation: the application of dissolved mineral nutrients via irrigation water.

Fertilizer: a substance, such as manure or chemical mixture (e.g. N-P-K) used to make soil more fertile and with a balanced approach improves plant growth, fruit size and fruit quality.

Field capacity: the amount of water held in a soil after it has been saturated and then allowed to drain away the water not held by soil particles.

Flame thrower, flamer: a torch used for spot application of heat to kill weeds in lieu of chemical herbicides, usually propane-fueled.

Floricane: the second year canes (stems) which bear fruit and then die.

Floricane-fruiting (FF): cultivars of raspberries that fruit in the summer of their second year.

GLOBALGAP: a private sector body that sets voluntary standards for the certification of production processes of agricultural (including aquaculture) products around the globe. The GLOBALGAP standard is primarily designed to reassure consumers about how food is produced on the farm by minimizing detrimental environmental impacts. GLOBALGAP serves as a practical manual for good agricultural practice (GAP) anywhere in the world.

Hardening or **hardening off:** refers to the ability of the plant to withstand fluctuating stress such as wet to dry or cold to heat and/or sunlight to darkness; hardening off is a process to regain strength of plants by placing plants into an environment of sufficient heat, sunlight and water after they have been subjected to stress of low moisture, low sunlight and low temperature, as in shipping of plants in a box over several days.

Hazard analysis critical control points (HACCP): a management system in which food safety is addressed through the analysis and control of biological, chemical and physical hazards from raw material production, procurement and handling, to manufacturing, distribution and consumption of the finished product.

Heading: the cutting (pruning) of a single main cane (stem) (central leader) for the purpose of encouraging lateral branching; also referred to as tipping or pinching on primocanes of black raspberries or on floricanes and primocanes of red raspberries.

Herbaceous host indicators: species and selections of herbaceous plants including beans, cucumbers, tobacco and amaranths that are particularly sensitive to viruses. When young plants of these herbaceous plants are inoculated with plant material of raspberry they will react in a visible and predicted manner to the presence of known viruses.

High tunnels: unheated, plastic-covered structures that provide an intermediate level of environmental protection and control compared to open field conditions or heated greenhouses. They are also known as hoophouses or plastic tunnels.

Idaeobatus: subsection of the genus *Rubus* that contains raspberry species, including *R. idaeus*, *R. strigosus* and about 200 other species that are spread around the world in the northern hemisphere and into Africa, Australia and Oceania.

Irrigation: the watering or application of water to farmland by means of ditches (furrows), trickle (drip) systems and/or pipes, either overhead, big guns or large traveling applicators.

Leaf light interception: the amount of light received by leaves; it is measured with light meters and quantified as photosynthetic photon flux (PPF) or photosynthetically active radiation (PAR).

Liquid feed: refers to nutrients being supplied to plants in a liquid form as compared to a dry form.

Loam: soil that is composed of balanced amounts of the three particle sizes; typically this is 40% sand, 40% silt and 20% clay.

Long day plant: a plant that flowers only after receiving illumination longer than a critical photoperiod, which is generally 11 to 14 h, depending on the species.

Manure tea: the liquid obtained when manures are soaked in a large volume of water; used as a source of dissolved mineral nutrients.

Marketable yield: the amount of product (fruit) that is sent to market; the total yield minus damaged, bruised or moldy berries equals marketable yield.

Mediterranean climate: climate that is characterized by warm to hot, dry summers and mild to cool, wet winters. Average temperatures are above 10°C (50°F) in their warmest months, and in the coldest months between 20°C (64°F) and −3°C (26.6°F).

Mosaic virus of raspberries: disease caused by a virus complex (more than one virus involved). Viruses of the mosaic complex cause the greatest reduction in growth, vigor, fruit yield and quality of any of the bramble viruses.

Municipal solid waste: garbage collected by community sanitation services. The biodegradable portion of this waste is sometimes composted and made available for farms and gardens.

Murashige and Skoog (MS) medium: the most commonly used plant growth medium for plant cell culture. MS medium was invented by plant scientists Toshio Murashige and Folke K. Skoog in 1962.

Nematodes: microscopic, unsegmented roundworms found worldwide in water, soil, decaying organic matter, plants and animals.

Nutraceutical: foodstuff, as a fortified food or a dietary supplement, that provides health or medical benefits in addition to its basic nutritional value.

Nutrient: a nutritive substance or ingredient; nutriment or something that promotes growth and repairs the natural wastage of organic life.

ORAC: oxygen radical absorbency capacity; a method of measuring antioxidant capacities in biological samples *in vivo*.

Permanent wilting point: the point at which remaining soil water is not available as it is held too tightly for the plant to extract it.

Phenology: the scientific study of periodic biological phenomena, such as growth and flowering in relation to climatic conditions.

Photosynthetically active radiation (PAR): the amount of light available for photosynthesis, which is light in the 400–700 nm wavelength range. PAR changes seasonally and varies depending on the latitude and time of day.

Phytonutrients: nutrients derived from plant material that have been shown to be necessary for sustaining human life.

***Phytophthora*:** any of the *Phytophthora* spp. water molds that cause root rots in wet soils. The main species of concern for raspberry growers is *Phytophthora rubi* (formerly *Phytophthora fragariae* var. *rubi*).

Phytophthora root rot: a disease caused by *Phytophthora fragariae* var. *rubi*, otherwise known as *Phytophthora rubi* (Wilcox and Duncan) Man in 't Veld, which limits production and longevity in raspberries in many of the areas in the world where raspberries are cultivated.

Plant available water (PAW): the amount of water available in a soil between field capacity and the permanent wilting point.

Precipitate: a substance separated from a solution or suspension by chemical or physical change.

Precipitation: in a mineral, the process of forming a precipitate; a deposit on the earth, as hail, mist, rain, sleet or snow.

Pre-emergent: a herbicide that kills weed seeds as they germinate; usually applied to the soil surface just before the weeds would normally appear.

Primocane: new cane (first year stem) on a raspberry plant that will flower and fruit the following year (floricane-fruiting type), or fruit from the tip basipetally in the current year (primocane fruiting).

Primocane fruiting (PF): fruiting on the first year canes, also called autumn-fruiting, everbearing, fall-bearing or remontant.

Primocane suppression: the practice of removing some or all of the primocanes during the early portion of the growing season.

Raspberry breeding program: program for the development of new cultivars of raspberries by the use of hybridization, growing of seedlings and selection of elite lines for commercial development.

Raspberry bushy dwarf virus (RBDV): a pollen- and seed-borne virus that is commonly found in red and black raspberry.

Rogue: selective removal of canes (stems) to the ground, usually in reference to those that emerge outside of the bed.

***Rubus coreanus*:** a black raspberry species found in Japan, Korea and north-eastern China. This species is very spiny and released cultivars have not been widely adopted as they are not very grower friendly.

***Rubus crataegifolius*:** a widespread red raspberry species found in Japan, Korea, China and the eastern part of the Russian Federation; the species is cultivated in its own right and harvested extensively from the wild in north-eastern China.

***Rubus idaeus*:** naturally occurring European wild species of red raspberry that is endemic or naturalized from Scotland through Europe and the Russian Federation into Asia; also known as or *Rubus idaeus* var. *vulgatus* or *Rubus idaeus* ssp. *vulgates*.

***Rubus neglectus* Peck:** naturally occurring or artificial hybrids between red and black raspberries.

***Rubus occidentalis* L.:** the black raspberry; a species native to the eastern USA and Canada; grown and cultivated in North America and in South Korea.

***Rubus strigosus*:** naturally occurring North American wild species of red raspberry; endemic or naturalized from Alaska through Canada from west to east and through the USA to northern Mexico; also known as *Rubus idaeus* var. strigosus or *Rubus idaeus* ssp. *strigosus*.

Sand: the largest-sized soil particles, easily visible to the naked eye, 0.05–2.0 mm in diameter.

Scald (in raspberries): results from overheating of fruit, even in the shade, and results in cell death; fruit turns a muddy red color and collapses.

Shelterbelt: similar to a windbreak; a row or multiple rows of shrubs and trees that decrease the force of oncoming winds. Shelterbelts are also used to reduce soil erosion, provide pollinator habitat and reduce drift of chemicals.

Silt: medium-sized soil particles, measuring 0.002–0.05 mm in diameter; roughly analogous to particles of white flour.

Soil drench: a pesticide applied in liquid form directly to the soil.

Sunburn (in raspberries): causes white drupelet disorder; the drupelet contents are overheated and the drupelet dies. This only occurs in direct sunlight. Similar symptoms may occur in some regions due to insect feeding.

Tipping: practice of removing the end portion of a raspberry cane (stem) to promote lateral branching; also referred to as pinching, particularly in black raspberries. See **Heading**.

Torus: receptacle of the raspberry fruit, which remains on the plant when the fruit is removed.

Trueness to type: a plant that has been examined and shown to be free of abnormal or atypical morphology. Trueness to type may be shown by DNA fingerprinting, where another genotype will quickly be distinguished from the desired clone.

Venturi: an instrument used in fertigation to inject chemicals or fertilizers into the irrigation system by using a constriction to create suction. It is named for physicist G.B. Venturi.

Verticillium wilt: one of the most serious diseases of raspberries; caused by a soil-borne fungus and causes wilting, stunting and eventually the death of a fruiting cane or the entire plant. The disease is usually more severe in black than in red raspberries.

Water stress: physiological stress that occurs when water is not easily accessible to roots.

White drupelet disorder: sunburn; caused by direct sunlight where the drupelet contents are overheated and the drupelet dies; similar symptoms occur in some regions due to insects (such as the stink bug) feeding on the fruit.

Wick and wiper (Wick wiper): a hand-held device used for spot treatment of herbicide, usually glyphosate.

Windbreak: a barrier composed of trees and shrubs or artificial materials to lessen the force of oncoming winds.

Yield: total amount of product (fruit) that is produced on a unit of land, quantified in kg/ha or lb/acre.

Glossary 2: Business Terms

Agribusiness: business activities engaged in by agriculture and farm production.
Agriculture: the production of livestock and/or crops; farming.
Agritourism: agricultural operations that bring visitors to the farm for educational or leisure activities, i.e. hayrides, corn mazes, pick-your-own berries, apple picking, horseback riding and wine tasting.
Asset to debt ratio: total assets divided by total liabilities; shows the proportion of a firm's assets that are financed through debt.
Assets: tangible or intangible economic resources stated on the balance sheet showing the value, i.e. cash on hand or in the bank, growing crops, crops in storage, livestock, etc.
Business: the activity of providing goods and services recorded by financial transactions.
Capital: the total finance required in order to operate a farm business.
Cash: money in any form, as coins, notes, checks, money orders or credit, that is immediately available.
Cash crop: crop produced to be sold for profit.
Cash flow: total amount of money in-flows and out-goes from a business.
Collective farm: a large farm run by a group of people (families) working together under government supervision for prices received and quantities produced. Generally, each family is given a house and plot of land for their own consumption.
Commercial: agricultural business with the intended purpose to make a profit.
Community (village organized) farming: arable fields, pastures and woods held in common, where each family has the right to produce food for the family and others in the village and were a major part of Europe and Russia in feudal times in the 19th and early 20th century.
Community-supported agriculture (CSA): an important social invention in industrialized countries starting in the USA in the 1980s and defined as where non-farm public 'members' pay for food in advance and receive a share of the food produced by the farm at harvest.
Cooperative: a jointly owned enterprise that produces and distributes goods and services, run for the benefit of its owners.
Cooperative credit: credit unions or banks formed by groups of people for the benefit of those using them; these have been shown to benefit families with low incomes on small tracts of land.

Credit: amount of money available for an individual or business account to borrow from a bank.
Currency: the type of money used in any particular country.
Debt: an amount that is owed or due.
DIRTI 5: a method used to determine fixed costs of machinery, irrigation, etc., which includes depreciation, interest, repairs (including fuel, maintenance), taxes and insurance (for equipment, shelter) or DIRTI.
Economics: a social science concerned chiefly with the description and analysis of production, distribution and consumption of goods and services.
Economies of size: related to the efficiencies of larger machines (technology) and the fully employed operator/laborer; average cost of production per unit declines as the size of operation grows.
Enterprise: a business organization, company.
Family farm: refers to farms owned and operated by a family, generally passed down from generation to generation.
Farm: a tract of land devoted to agricultural purposes, such as crop production or raising livestock animals.
Farmer: a person who cultivates land to produces crops or raises livestock.
Farmers' market: customers come to a centralized location where farmers offer their farm products for sale.
Farm management: the decision-making process of a farm business in which the resources are utilized wisely or an applied science dealing with the biology, economics, technology and the social aspects to achieve a return on investment to land, management and labor, including farm markets and community-supported agriculture, where the public interacts and connects with the land.
Farmstead: a farm and the adjacent building area, which may include a farmhouse for the owner or tenant, barns, housing for livestock, with a garden and orchard near the house.
Finance: to raise or provide funds or capital.
Financial risk: any financial situation which jeopardizes variability of the farm and involves the proportion of debt and equity in the entire farm firm or current assets and current liabilities.
Garden: a plot of land where herbs, fruits, flowers or vegetables are cultivated.
Gross farm income: the total cash income from the sale of farm products, such as lumber, livestock, crops and/or machinery.
Human capital: skills, knowledge and experience possessed by an individual allowing the capacity to do valuable work to produce goods and services and interact in society.
Income: financial earnings or benefit, measured in money received from capital or labor.
Increasing risk principle: the greater the percent of borrowed assets in total farm assets (leverage), the greater the risk of becoming insolvent or illiquid.
Internal rate of return: a discounted cash flow (DCF) method used to compare two or more investments with different costs and revenue streams to calculate a rate of return of cash flow across time. It takes into account the 'time value of money' economic principle.
Labor: expenditure of physical work or mental effort by an individual which provides goods and services in an economy.

Labor efficiency: the amount of input (labor hours) required to produce an amount of output (kilograms or pounds harvested, etc.), expressed as a ratio, i.e. kilograms or pounds of raspberries harvested per hour.
Law of diminishing return: the economic principle in a production function which states that while additional units of input increase output, it occurs at a declining incremental (marginal) rate.
Liquidity: the ability of the farm manager to meet short-term debt as it becomes due; debt paying ability.
Long run: the time horizon beyond one production season, as for grain, but can be 5–15 years for fruit crops; where many costs are considered variable. Land costs should be entered as a cost (either lease, rent or purchased) or where family labor may increase or decrease.
Manual labor: physical work done with the hands; labor accomplished by people manually, such as in manual pruning, manual planting or harvesting by hand harvest.
Market: a meeting together of people for the purpose of trade by private purpose and sale (not an auction).
Marketable: fit to be offered for sale in a market; saleable.
Marketable yield: only those items that are most saleable or berries that are not bruised, moldy or of poor color.
Mechanical: of or relating to machinery or tools, such as mechanical pruning, mechanical harvest or mechanical planting.
Net farm income: net cash income adjusted for inventory increases or decreases and depreciation.
Net present value: a discounted cash flow (DCF) method which is the sum of adjusted dollars over a period of years (cash flow); can indicate the ability to pay interest on the investment over time from future returns and also assess risk. An interest rate is selected for this procedure.
Net return: income less expenses; the remaining funds when subtracting costs from returns.
Net worth statement: the listing of all assets (values of land, buildings, equipment, etc.) minus all liabilities (debt as loans, mortgage and payments due, etc.).
Peasant: one of many persons tilling the soil, as small landowners or as labors; generally those who are in the lower socioeconomic class in Europe.
Pick-your-own (PYO): the public are allowed to enter a commercial farm to pick (harvest) fruit or vegetable crops and pay for the amount (volume or weight) that is harvested.
Profit system: free enterprise.
Profit: the excess of returns over expenditures (costs) in a transaction or series of transactions; the excess of the selling price of goods over their cost of goods.
Rate of return on the investment: net farm income less unpaid farm labor and operator's labor divided by the total value of the farm investment.
Return: money received from an investment or yield on investment.
Revenue: income that comes back from an investment or the total income produced by a given source.
Roadside market: generally refers to a seasonal farm market located by a road or highway.
Rural life: life in the countryside (non-urban area) or small town in the countryside.

Serf: a peasant who cannot legally leave the land on which he or she works.

Short run: the time horizon of less than 1 year for grain crops; 1–5 years for fruit crops and where many inputs are considered fixed.

Subsistence farming: providing only the basic needs of living from food, fiber and forestry products and/or utilizing the production and processing of food, fiber and lumber to sustain life in rural life.

Technology: the use of science for practical purposes, such as substitution of machinery for labor; computers and programs for accounting, management and fruit traceability, social media for advertising, genetically improved plants; chemical pest control; and the efficient monitoring and use of lime and fertilizer.

Three 'R's' of credit: *risk* (financial, ability to obtain a loan based on net worth), *returns* (net farm income), *repayment* (of loans).

Time value of money: money is worth more today than tomorrow (future date) usually due to uncertainty, alternative uses and/or inflation.

Traceability: a system or process that traces or tracks the fruit from production and harvest through to the customer.

Urban: in or around a town or city.

Value: a fair return or equivalent in goods, services or money for something exchanged.

Value added: increasing the value of a commodity by completing a process that creates additional value, such as producing jam, pastries or wine from raspberries and selling the new product at a higher return.

Value chain: steps that a business (farmer) goes through from the beginning step of production through the final step of purchase by the customer, as the end user.

Wages: money paid to workers, usually on an hourly basis.

INDEX

anthocyanin 30
anthracnose (*Elsinoe veneta*) 81,
    142–143, 171
antioxidants 29–30
aphids 11, 68, 148–149
    large raspberry aphid (*Amphorophora agathonica*) 11, 148
    small (leaf curling) raspberry aphid (*Amphorophora idaei*) 148–149
    virus transmission 137, 148
arabis mosaic virus 68
Asian species 14–15
Australia 225–226

bacterial crown gall (*Agrobacterium tumefaciens*) 136
black raspberry 9–12, 55, 66, 84
    pruning 127–130
black raspberry necrosis virus (BRNV) 11
boron 95–96
Botrytis cinerea 41, 80–81, 143–145, 171
breeding 5–7, 8–9, 10–12, 16, 55, 70–71
    goals 57–71
        climatic adaptation 68–69
        disease resistance 67–68, 136
        flavor 66–67
        fruit color 57–66

fruit quality 57
growth habit 70
insect resistance 67–68
shelf life 67
spine-free canes 70
programs 56–57
bud break 35–38

Canada 222–224
cane
    blight (*Kalmusia coniothyrium*) 81, 142
    borer (*Oberea bimaculata*) 146–147
    Botrytis 143
    diseases 80–81, 170, 171
    frost damage 36–37, 46, 47, 145
    gall (*Agrobacterium rubi*) 136
    growth habit 70
    maggot (*Pegomya rubivora*) 147–148
    morphology 23–24
    spine-free 70
    spot (*Elsinoe veneta*) 81, 142–143
    support 99
    wind damage 40
cation exchange capacity (CEC) 92, 104–105
certification schemes 73–74
China 226
classification 2–3
clean culture system 104

273

climate 29, 33, 46–47, 157
  change 43
  fall (autumn) 39–40
  frost damage 36–37
  humidity 41–42
  rainfall 39, 41
  spring 37–38
  summer 38–39
  sunlight 29, 42–43
  temperature 28–29, 30, 33–40, 47
  wind 29, 40
  winter 35–37
CO_2 187–188
cold hardiness 27, 34–35, 36–37, 68–69
community-supported agriculture (CSA) 191, 193, 194
concentrate 197
cooling 182–185
  calculation of cooling time 185–186
cover crops 93, 107
crop production 157–158
  berry quality 38–39, 57, 161–162
  fertilization 105–106, 167–168, 170
  floricane-fruiting cultivars 26–28, 122–124, 127–130, 160
  global production 213–214
    Australia 225–226
    Canada 222–224
    China 226
    Europe 216–222
    fresh market cultivars 215–216
    marketing drivers 227–233
    mechanical harvesting 215
    Mexico 222–224
    New Zealand 225–226
    South Korea 226
    southern hemisphere 227
    USA 222–224
  plant health 159–160
  primocane suppression 169–170
  primocane-fruiting cultivars 28–29, 127–130, 161
  protected crops 43
  pruning 28, 69, 121–122, 147, 171
    black raspberry 127–130
    protected crops 130–131
    purple raspberry 130
    red raspberry 122–124, 127
  water management 103, 109, 158, 162–165
    evapotranspiration (ET) 163–164
    irrigation 41, 100, 106, 108, 109–111, 112–115, 163–165, 171–172
    plant available water (PAW) 112–113
    requirements 111–113
  weed management 97, 100, 104, 115–117, 152–153, 166–167, 172
  see also crop protection
crop protection 134, 154–155, 169, 170–171
  aphid resistance 11, 68, 148–149
  application 153–154
  biological control 150
  cover crops 93
  disease free stock *see* nursery production
  disease resistance 67–68, 136
  frost damage 36–37, 46, 47, 145
  fumigation 93, 133
  fungicides 139, 141, 143, 171
  grey mold 41, 144–145
  insect resistance 67–68
  insecticides 146, 147, 148, 149–150, 151, 170–171
  miticides 151
  natural predators 151
  plant passport 74
  virus resistance 68
  wind damage 40
cropping period 43
  elevation 47–48
crown borer (*Pennisetia marginata*) 146
cultivars 58–65, 215–216, 228–230
  'Anne' 66, 67, 68
  'Autumn Bliss' 29
  'Boyne' 68, 69
  'Caroline' 66, 68, 69, 70
  'Crimson Giant' 57, 67, 69
  'Encore' 57, 67, 69
  'Glen Ample' 57, 67
  'Glen Lyon' 57
  'Heritage' 29, 30, 66, 67, 69, 70, 139

'Himbo Top' 57, 67
'Jewel' 66
'Killarney' 68, 69
'Latham' 68, 69
'Maravilla' 57, 67, 69
'Meeker' 66, 67, 70, 215
'Munger' 66
selection 178–179
'Summit' 35
'Tulameen' 39–40, 67
'Willamette' 40, 68, 136, 215
see also breeding

disease 133–134, 171
  anthracnose (*Elsinoe veneta*) 81, 142–143, 171
  arabis mosaic virus 68
  bacterial crown gall (*Agrobacterium tumefaciens*) 136
  black raspberry necrosis virus (BRNV) 11
  Botrytis cinerea 41, 80–81, 143–145, 171
  cane blight (*Kalmusia coniothyrium*) 81, 142
  cane gall (*Agrobacterium rubi*) 136
  cane spot (*Elsinoe veneta*) 81, 142–143
  downy mildew 80
  grey mold 41, 80–81, 143–145, 171
  late leaf rust (*Puccinastrum* spp.) 139–141
  leaf spot (*Sphaerulina rubi*) 141
  nematodes 76, 93
    virus transmission 138–139
  orange rust 80, 139
  Phytophthora root rot 67, 68, 76, 78–80, 108–109, 135–136
  plant health testing 76–77
    cane diseases 80–81
    leaf diseases 81–82
    root diseases 78–80
    viruses 77–78
  powdery mildew (*Sphaerotheca macularis*) 82, 141–142
  raspberry bushy dwarf virus (RBDV) 67, 68, 76, 77, 138, 225–226
  raspberry ringspot virus 68, 138–139
  Rubus yellow net virus (RYNV) 137
  soil-borne 76
  spur blight (*Didymella applanata*) 81, 143
  tomato black ring virus 68
  Verticillium wilt 68, 79–80, 134–135
  viruses 77–78, 137
  yellow rust (*Phragmidium rubi-idaei*) 82, 122, 141
dormancy 26–27, 35–36
downy mildew (*Peronospora sparsa*) 80
drip irrigation 43, 106, 109–110, 112, 113, 136, 144, 163, 245, 246, 258
drought 41

economics 201–203, 211
  enterprise budget 203–204
  equipment 206
  internal rate of return (IRR) 205, 209–210
  labor 206–207, 208–209
  lifetime costs 207–209
  machine harvesting 211
  net present value (NPV) 205, 209–210
  productivity 211
  risk management 204–205
  whole farm management 203
elevation 47–48
ethylene 179
Europe
  production 216–222
  Rubus species 13–14
evapotranspiration (ET) 163–164

fall (autumn) 39–40
farm management 201–203, 211
  enterprise budget 203–204
  equipment 206
  internal rate of return (IRR) 205, 209–210
  labor 206–207, 208–209
  lifetime costs 207–209
  machine harvesting 211
  net present value (NPV) 205, 209–210

farm management *continued*
  productivity 211
  risk management 204–205
  whole farm management 203
fertigation 43, 106, 115, 163, 170, 245, 246, 253, 255, 258–260
fertilization 105–106, 167–168, 170
floricane-fruiting cultivars 58–61
  growth 26–28, 160
  pruning 122–124, 127–130
  training 124–129
flower initiation 39–40
frost damage 36–37, 46, 47, 145
fruit 160, 188
  color 57–66
  ethylene 179
  flavor 66–67
  grey mold 144–145, 171
  harvesting
    fresh market 179–180
    mechanical 130, 180
    packaging 181–182
    processing 180
    sanitation 180–181
    training 180
  marketing 52–53, 191–192
    community-supported agriculture (CSA) 191, 193, 194
    cultivar selection 178–179
    direct to consumer 193–195
    drivers 227–233
    pick-your-own (PYO) 191, 192–193, 194–195
    price 195
    wholesale 195–198
  morphology 24–25
  postharvest 27–28, 50–51
  processing 51–52, 180, 231–232
  quality 38–39, 57, 161–162, 179–180
  rot 41
  shelf life 67, 182–183, 187–188
  storage 182–185
    calculation of cooling time 185–186
    modified atmosphere 187–188
  transportation 49, 89, 186–187
  white drupelet disorder 43, 145–146

fruitworms (*Byturus* spp.) 150–151
fumigation 93, 133
fungicides 139, 141, 143, 171

geographic distribution 3, 33–34
  Asian species 14–15
  European species 13–14
  Oceania species 15
  North American species 15
global production 213–214
  Australia 225–226
  Canada 222–224
  China 226
  Europe 216–222
  fresh market cultivars 215–216
  marketing drivers
    commercial fresh market 230–231
    cultivar quality 228–230
    nutritional value 227–228
    on-farm sales 230
    processed fruit 231–232
    value-added products 232–233
    year-round production 228
  mechanical harvesting 215
  Mexico 222–224
  New Zealand 225–226
  South Korea 226
  southern hemisphere 227
  USA 222–224
green June beetle (*Cotinis nitida*) 149
grey mold (*Botrytis cinerea*) 41, 80–81, 143–145, 171
growth
  floricane-fruiting cultivars 26–28, 122–124, 127–130, 160
  primocane-fruiting cultivars 28–29, 127–130, 161
  sunlight 42

harvesting 188
  fresh market 179–180
  mechanical 130, 180
  packaging 181–182
  processing 180
  sanitation 180–181

storage 182–185
    calculation of cooling time
        185–186
    training 180
herbicide strip with grass (sod) culture
        104–107, 169
    endophytes 107
    grass selection 106–107
    water management 112–113
herbicides 97, 104, 115–117, 152–153,
        166–167, 172
history 1–2
humidity 41–42

in vitro propagation 75–76
individually quick frozen (IQF)
        raspberries 52, 185, 197,
        231–232
insecticides 146, 147, 148, 149–150,
        151, 170–171
insects 133–134, 146, 170–171
    aphids 11, 68, 148–149
        large raspberry aphid (*Amphorophora
            agathonica*) 11, 148
        small (leaf curling) raspberry aphid
            (*Amphorophora idaei*) 148–149
        virus transmission 137, 148
    fruitworms (*Byturus* spp.) 150–151
    green June beetle (*Cotinis nitida*) 149
    Japanese beetles (*Popillia japonica*)
        149
    New Zealand grass grub (*Costelytra
        zealandica*) 149–150
    picnic beetle (*Glischrochilus* spp.) 150
    potato leaf hopper (*Empoasca fabae*)
        67–68, 152
    raspberry cane borer (*Oberea
        bimaculata*) 146–147
    raspberry cane maggot (*Pegomya
        rubivora*) 147–148
    raspberry cane midge (*Resseliella
        theobaldi*) 148
    raspberry crown borer (*Pennisetia
        marginata*) 146
    raspberry leaf and bud mite 81
    raspberry sawfly (*Monophadnoides
        geniculatus*) 150

    red-necked cane borer (*Agrilus rificollis*)
        147
    rose chafer (*Macrodactylus subspinosus*)
        149
    strawberry clipper (*Anthonomus
        signatus*) 147
    tarnished plant bug (*Lygus lineolaris*)
        151
    tree crickets 148
    two-spotted spider mite (*Tetranychus
        urticae*) 151
    yellowjackets 152
irrigation 41, 100, 106, 108, 112–113,
        163–165, 171–172
    business plan 115
    monitoring 113–115
    quality 110–111
    water supply 109–110
Italian production 220

Japanese beetles (*Popillia japonica*) 149

labor 50, 206–207, 208–209
large raspberry aphid (*Aphis agathonica*)
        11, 148
late leaf rust (*Puccinastrum* spp.)
        139–141
leaf spot (*Sphaerulina rubi*) 141
life cycle 7–8
lime 94–95

management 201–203, 211
    enterprise budget 203–204
    equipment 206
    internal rate of return (IRR) 205,
        209–210
    labor 206–207, 208–209
    lifetime costs 207–209
    machine harvesting 211
    net present value (NPV) 205,
        209–210
    productivity 211
    risk management 204–205
    whole farm management 203

marketing 52–53, 191–192
  community-supported agriculture (CSA) 191, 193, 194
  cultivar selection 178–179
  direct to consumer 193–195
  drivers
    commercial fresh market 230–231
    cultivar quality 228–230
    nutritional value 227–228
    on-farm sales 230
    processed fruit 231–232
    value-added products 232–233
    year-round production 228
  pick-your-own (PYO) 191, 192–193, 194–195
  price 195
  wholesale 195–198
mechanical harvesting 130, 180, 211, 215
medicinal properties 1–2, 29–30
Mexico 222–224
miticides 151
modified atmosphere 187–188
molecular markers 11–12
morphology
  canes 23–24
  fruit 24–25
  root 21–22
mulching 100

natural predators 151
nematodes 76, 93
  virus transmission 138–139
New Zealand 225–226
nitrogen 95, 96, 170
North American species 15
nuclear stock 74
nursery production 73–74
  certification schemes 73–74
  field propagation 76
  in vitro propagation 75–76
  plant health testing 76–77
    cane diseases 80–81
    leaf diseases 81–82
    root diseases 78–80
    viruses 77–78
  trueness to type testing 77

nutrients 94–96
  cation exchange capacity (CEC) 92, 104–105
  fertilization 105–106, 167–168, 170
nutritional value 227–228

orange rust (*Arthuriomyces peckianus*) 80, 139
organic matter 92, 93

Peronospora sparsa see downy mildew
pesticide application 153–154
pests 133–134, 146, 170–171
  aphids 11, 68, 148–149
    large raspberry aphid (*Amphorophora agathonica*) 11, 148
    small (leaf curling) raspberry aphid (*Amphorophora idaei*) 148–149
    virus transmission 137, 148
  fruitworms (*Byturus* spp.) 150–151
  green June beetle (*Cotinis nitida*) 149
  Japanese beetles (*Popillia japonica*) 149
  nematodes 76
  New Zealand grass grub (*Costelytra zealandica*) 149–150
  picnic beetle (*Glischrochilus* spp.) 150
  potato leaf hopper (*Empoasca fabae*) 67–68, 152
  raspberry cane borer (*Oberea bimaculata*) 146–147
  raspberry cane maggot (*Pegomya rubivora*) 147–148
  raspberry cane midge (*Resseliella theobaldi*) 148
  raspberry crown borer (*Pennisetia marginata*) 146
  raspberry leaf and bud mite 81
  raspberry sawfly (*Monophadnoides geniculatus*) 150
  red-necked cane borer (*Agrilus rificollis*) 147
  rose chafer (*Macrodactylus subspinosus*) 149
  strawberry clipper (*Anthonomus signatus*) 147

tarnished plant bug (*Lygus lineolaris*) 151
tree crickets 148
two-spotted spider mite (*Tetranychus urticae*) 151
yellow jackets 152
pH 94–95, 105
phosphorus 95
physiology 25–26
  climate 29, 33, 46–47, 157
    change 43
    fall (autumn) 39–40
    frost damage 36–37
    humidity 41–42
    rainfall 39, 41
    spring 37–38
    summer 38–39
    sunlight 29, 42–43
    temperature 28–29, 30, 33–40, 47
    wind 29, 40
    winter 35–37
  growth
    floricane-fruiting cultivars 26–28, 122–124, 127–130, 160
    primocane-fruiting cultivars 28–29, 127–130, 161
    sunlight 42
Phytophthora root rot 67, 68, 76, 78–80, 108–109, 135–136
pick-your-own (PYO) 191, 192–193, 194–195, 203–204
picnic beetle (*Glischrochilus* spp.) 150
plant available water (PAW) 112–113
plant health testing 76–77
  cane diseases 80–81
  leaf diseases 81–82
  root diseases 78–80
  viruses 77–78
plant passport 74
plant propagation 83–84
  dormant plants 87–88
  nursery production 73–74
    certification schemes 73–74
    field propagation 76
    in vitro propagation 75–76
    plant health testing 76–82
    trueness to type testing 77
  planting 89, 99–100

storage 88
tissue culture 75–76, 84–87
planting 89, 99–100
  density 125–126
pollination 27
Portuguese production 220
postharvest 27–28, 50–51, 177–178, 188
  harvesting
    fresh market 179–180
    mechanical 130, 180
    packaging 181–182
    processing 180
    sanitation 180–181
    training 180
  marketing 52–53, 191–192
    community-supported agriculture (CSA) 191, 193, 194
    cultivar selection 178–179
    direct to consumer 193–195
    drivers 227–233
    pick-your-own (PYO) 191, 192–193, 194–195
    price 195
    wholesale 195–198
  processing 51–52, 180
  quality 38–39, 57, 161–162, 179–180
  shelf life 67, 182–183, 187–188
  storage 182–185
    calculation of cooling time 185–186
    modified atmosphere 187–188
  transportation 49, 89, 186–187
potassium 95
potato leaf hopper (*Empoasca fabae*) 67–68, 152
powdery mildew (*Sphaerotheca macularis*) 82, 123, 141–142
primocane-fruiting cultivars 8–9, 62–65
  growth 28–29, 161
  pruning 127–130
processed fruit 231–232
processing plants 51–52
production 157–158
  berry quality 38–39, 57, 161–162
  fertilization 105–106, 167–168, 170
  floricane-fruiting cultivars 26–28, 122–124, 127–130, 160

production *continued*
　global production 213–214
　　Australia 225–226
　　Canada 222–224
　　China 226
　　Europe 216–222
　　fresh market cultivars 215–216
　　marketing drivers 227–233
　　mechanical harvesting 215
　　Mexico 222–224
　　New Zealand 225–226
　　South Korea 226
　　southern hemisphere 227
　　USA 222–224
　plant health 159–160
　primocane-fruiting cultivars 28–29, 127–130, 161
　primocane suppression 169–170
　protected crops 43
　pruning 28, 69, 121–122, 147, 171
　　black raspberry 127–130
　　protected crops 130–131
　　purple raspberry 130
　　red raspberry 122–124, 127
　water management 103, 109, 158, 162–165
　　evapotranspiration (ET) 163–164
　　irrigation 41, 100, 106, 108, 109–111, 112–115, 163–165, 171–172
　　plant available water (PAW) 112–113
　　requirements 111–113
　weed management 97, 100, 104, 115–117, 152–153, 166–167, 172
　see also crop protection
protected crops 107–108, 130–131
pruning 28, 69, 121–122, 147, 171
　black raspberry 127–130
　protected crops 130–131
　purple raspberry 130
　red raspberry
　　floricanes 122–124
　　primocanes 127
purple raspberry 12–13, 66, 84
　pruning 130

rainfall 39, 41
raised beds 104, 108–109
　water management 112–113
raspberry 21
　bushy dwarf virus (RBDV) 67, 68, 76, 77, 138, 225–226
　cane borer (*Oberea bimaculata*) 146–147
　cane maggot (*Pegomya rubivora*) 147–148
　cane midge (*Resseliella theobaldi*) 148
　crown borer (*Pennisetia marginata*) 146
　fruitworms (*Byturus* spp.) 150–151
　history of 1–2, 213
　leaf and bud mite 81
　ringspot virus 68, 138–139
　sawfly (*Monophadnoides geniculatus*) 150
　yellow rust (*Phragmidium rubi-idaei*) 82, 122
red-necked cane borer (*Agrilus rificollis*) 147
red raspberry 5–7, 55, 57–66, 83–84
　pruning
　　floricanes 122–124
　　primocanes 127
root
　borer (*Pennisetia marginata*) 146
　diseases 78–80, 108–109
　morphology 21–22
rose chafer (*Macrodactylus subspinosus*) 149
Rubus
　Asian species 14–15
　European species 13–14
　idaeus 5–7, 12
　leucodermis 9–12
　North American species 15
　occidentalis 9–12
　Oceania species 15
Rubus yellow net virus (RYNV) 137

sawfly (*Monophadnoides geniculatus*) 150
scarab beetles 149–150
shelf life 67, 182–183, 187–188

site
  layout 98–99
  preparation 91, 97–98
  selection
    access 48
    climate 46–47
    elevation 47–48
    labor 50
    location 48–49
    market 52–53
    packing sheds 50–51
    processing plants 51–52
    soil 45
    topography 46, 98
    transportation 49
small (leaf curling) raspberry aphid
    (*Amphorophora idaei*) 148–149
soil 45, 158
  biological properties 92–93
  cation exchange capacity (CEC) 92,
    104–105
  chemical properties 94–97
  management 103, 104–109
    clean culture system 104
    fertilization 105–106
    herbicide strip with grass (sod)
      culture 104–107, 112–113
    inter-row cover crop 107
    raised beds 104, 108–109,
      112–113
  pH 94–95, 105
  physical properties 91–92
  preparation 97–98
  structure 104, 108
  testing 94, 97, 167
  water-holding capacity 111–112
South Korea 226
Spanish production 220
spray applications 153–154
spray equipment selection 153–154
spring 37–38
spur blight (*Didymella applanata*) 81, 143
storage 182–185
  calculation of cooling time 185–186
  modified atmosphere 187–188
strawberry clipper (*Anthonomus signatus*)
    147

summer 38–39
sunburn 38–39
sunlight 29, 42–43

tarnished plant bug (*Lygus lineolaris*)
    151
temperature 28–29, 30, 33–40, 47
  high temperature resistance 69
tissue culture 75–76, 84–87
  media 85–86
  rooting 86–87
  sterilization 85
tomato black ring virus 68
topography 46, 98
traceability 197–198
training 121–122
  trellis 99, 121, 124–127, 127–129,
    130–131, 158
transportation 49, 89, 186–187
tree crickets 148
trellis 99, 121, 124–127, 127–129,
    130–131, 158
trueness to type testing 77
tunnels 107–108
two-spotted spider mite (*Tetranychus
    urticae*) 151

UK production 216–219
USA 222–224

Verticillium wilt 68, 79–80, 134–135
viruses 77–78, 137
  arabis mosaic virus 68
  black raspberry necrosis virus (BRNV)
    11
  nematode virus transmission
    138–139
  raspberry bushy dwarf virus (RBDV)
    67, 68, 76, 77, 138, 225–226
  raspberry ringspot virus 68,
    138–139
  Rubus yellow net virus (RYNV) 137
  tomato black ring virus 68
vitamin content 30

water management 103, 109, 158, 162–165
   evapotranspiration (ET) 163–164
   irrigation 41, 100, 106, 108, 112–113, 163–165, 171–172
      business plan 115
      monitoring 113–115
      quality 110–111
      supply 109–110
   plant available water (PAW) 112–113
   requirements 111–113
weather stations 114–115
weed management 97, 100, 104, 115–117, 152–153, 166–167, 172
white drupelet disorder 43, 145–146
wholesale market 195–198
wind 29, 40
windbreaks 40, 47, 98–99, 235–242
winter 35–37
world production 213–214
   Australia 225–226
   Canada 222–224
   China 226
   Europe 216–222
   fresh market cultivars 215–216
   marketing drivers
      commercial fresh market 230–231
      cultivar quality 228–230
      nutritional value 227–228
      on-farm sales 230
      processed fruit 231–232
      value-added products 232–233
      year-round production 228
   mechanical harvesting 215
   Mexico 222–224
   New Zealand 225–226
   South Korea 226
   southern hemisphere 227
   USA 222–224

year-round production 228
yellow raspberry 13, 66
yellow rust (*Phragmidium rubi-idaei*) 82, 122, 141
yellowjackets 152